Methods for Solving Complex Problems in Fluids Engineering

Can Kang · Haixia Liu · Yongchao Zhang
Ning Mao

Methods for Solving Complex Problems in Fluids Engineering

Science Press
Beijing

Springer

Can Kang
School of Energy and Power Engineering
Jiangsu University
Zhenjiang, Jiangsu, China

Yongchao Zhang
School of Energy and Power Engineering
Jiangsu University
Zhenjiang, Jiangsu, China

Haixia Liu
School of Materials Science and Engineering
Jiangsu University
Zhenjiang, Jiangsu, China

Ning Mao
School of Energy and Power Engineering
Jiangsu University
Zhenjiang, Jiangsu, China

ISBN 978-981-13-2648-6 ISBN 978-981-13-2649-3 (eBook)
https://doi.org/10.1007/978-981-13-2649-3

Jointly published with Science Press, Beijing, China

The print edition is not for sale in China Mainland. Customers from China Mainland please order the print book from: Science Press.

ISBN of the China Mainland edition: 978-7-03-058750-3

Library of Congress Control Number: 2018955165

This Springer imprint is published by the registered company Springer Nature Singapore Pte Ltd.
The registered company address is: 152 Beach Road, #21-01/04 Gateway East, Singapore 189721, Singapore

Preface

Fluids engineering covers a wide spectrum of topics from the miniature blood pump to the large-scale pump station. As a matter of fact, many problems encountered in fluids engineering cannot be explained with existing knowledge. For instance, regarding fluid machinery, an important constituent of the flow system, new problems arose constantly in recent years with the extension of its application. Some problems are challenging. This prompts researchers to seek advanced research techniques to recognize and explain the influential factors underlying the problems. Moreover, a bridge between academic achievements and engineering applications is sorely needed.

This book is intended to provide insight into problems encountered in fluids engineering. The emphasis is on the application of experimental and numerical techniques. For experimental techniques, nonintrusive optical measurement and flow visualization are emphasized. The numerical techniques are not limited to computational fluid dynamics (CFD), the application of the finite element method is discussed as well. The cases which the authors selected involve various flows such as rotating flows and multiphase flows; they are related not just to a single machine but also to a whole system. The readers might expand their vision or plan a further exploration based on some cases presented here.

Chapter 1 contains an overview of the complex problems encountered in fluids engineering, particularly, in recent years. Meanwhile, generally used methods for addressing these problems are briefly introduced. In Chap. 2, experimental techniques used in treating complex flows are introduced. Optical measurement and flow visualization techniques are emphasized. Numerical methods, which are sometimes equivalently important as experimental ones, are introduced as well. In Chap. 3, the waterjet with distinct features is investigated. High-pressure waterjet is issued from a small nozzle and flows near the tiny jet stream deserve a full consideration from every aspect. Cavitation might occur as the waterjet is injected into stationary water; therefore, both flow characteristics and cavitation phenomenon are analyzed in this chapter. Chapter 4 deals with the motions of the bubble in stationary water and in flowing water. Tracing transient bubble characteristics and bubble image processing are two aspects focused in this chapter. Ventilation serves as an effective measure for increasing the gas or oxygen content in liquid. In Chap. 5, an attempt is made to inject air into water through a cylinder. Gas bubbles trapped in such a unique wake flow are studied. Chapter 6 records a drag-type hydraulic rotor. Such a rotor can be traced back to the conventional Savonius rotor. In this chapter, the rotation of this rotor is driven by flowing water. Flow patterns near the rotor are observed. The impeller pump is used commonly in fluids engineering. Flows in the impeller pump remain a focus in relevant studies. In Chap. 7, flows in the impeller pump are investigated. It is anticipated to relate the flow patterns with the geometry of the pump hydraulic components. Meanwhile, the effects of operation conditions are taken into account. In Chap. 8, cavitation in the pump is discussed. Chapter 9 treats the interaction between flows and solid structures; specifically, the flow–structure interaction in the impeller pump is explained. In this book, the mechanism of complex flows is particularly emphasized. After all, the understanding of flow mechanisms is the base of the optimal design of relevant machinery and system.

The authors thank the support of National Natural Science Foundation of China (Grant Nos. 51676087, 51775251, 51376081, 51205171, and 50806031). The authors are also grateful for the financial support of Top-notch Academic Programs Project of Jiangsu Higher Education Institutions. The authors would like to thank senior engineers from Shanghai Marine Research Institute, and they are Bing Li, Weifeng Gong, Changjiang Li, Yanhua Feng, Kejing Ding, Wenbin Zhang, and Haifei Gu for their valuable comments on the relationship between flow characteristics in the pump and the pump performance.

Zhenjiang, China

Can Kang
Haixia Liu
Yongchao Zhang
Ning Mao

Contents

About the Authors

Can Kang is a Professor of fluids engineering at Jiangsu University of China. He is a senior member of Chinese Mechanical Engineering Society (CMES), a member of American Society of Mechanical Engineers (ASME), International Association for Hydro-Environment Engineering and Research (IAHR), and Chinese Society of Theoretical and Applied Mechanics (CSTAM). He has received eight honors and awards in fluid machinery and fluids engineering. He has authored and co-authored 6 monographs and textbooks and has published more than 120 academic papers. His current research interests include impeller pumps, submerged waterjet, cavitation, and flow visualization techniques.

Haixia Liu is a Professor of material science and engineering at Jiangsu University of China. She has performed extensive investigations on surface modification with cavitating technology since 2010. She is the author of 45 academic papers and the co-author of a monograph. Her current research interests are focused on cavitation erosion induced by ultrasonic cavitation and waterjet cavitation.

Yongchao Zhang is a Ph.D. candidate of fluids engineering at Jiangsu University of China. His research interests include pumps applied in nuclear engineering, liquid metal flows, and drag-type hydraulic rotors. He is performing studies using both experimental and numerical tools.

Ning Mao is a Ph.D. candidate of fluids engineering at Jiangsu University of China. His research interests include bubble dynamics and cavitation visualization. Particularly, he uses high-speed photography and in-house image-processing codes to obtain quantitative information of bubbles.

Brief Introduction

This book presents a comprehensive introduction to flow-related problems encountered in fluids engineering. Methods and strategies used for treating these problems are emphasized. The importance of experimental and numerical tools is substantiated through several cases. Although fluids engineering encompasses a wide range of subjects, the cases exhibited in this book are representative in today's view. Meanwhile, this book can be used as a material for an introductory graduate course on complex flow problems in fluids engineering.

Chapter 1
Introduction to Complex Problems in Fluids Engineering

Abstract Fluids engineering is a huge subject and involves not just fluid machinery but also the entire flow system. The major foundations of fluids engineering are fluid mechanics and mechanical principles. A tendency in recent years is new flow phenomena are constantly revealed. Meanwhile, the complexity of these flow phenomena is apparently high. A broad range of challenges in fluids engineering are being faced by both engineers and researchers. Fluid itself is inherently complex; meanwhile, operation environment nurtures many unknowns. In this chapter, a brief overview of the problems encountered in fluids engineering is presented. Fundamental flow characteristics and typical flow system are covered.

1.1 Background Knowledge

The American Engineers' Council for Professional Development (ECPD) has defined engineering as: The creative application of scientific principles to design or develop structures, machines, apparatus, or manufacturing processes, or works utilizing them singly or in combination; or to construct or operate the same with full cognizance of their design; or to forecast their behavior under specific operating conditions; all as respects an intended function, economics of operation and safety to life and property. Such a definition applies to all kinds of engineering and engineering activities. In many countries, fluids engineering is not an independent engineering branch, it is generally attached to mechanical engineering or chemical engineering.

Engineering applications exert significant restrictive conditions on flows. For instance, in the flow passages of the impeller pump, flows are subjected to various effects such as the flow rate, the blade geometry, the interaction between the impeller and the diffuser. In many occasions, the cavitation phenomenon has to be considered. These factors are sometimes inter-connected. The flow structure and its scale are two factors that are emphasized in fluids engineering. For the former, a general viewpoint argues that it depends on the restriction of the fluid-wetted wall. This is a macroscopic viewpoint. In fact, we do not know exactly what kind of solid

C. Kang et al., *Methods for Solving Complex Problems in Fluids Engineering*,
https://doi.org/10.1007/978-981-13-2649-3_1

wall would result in what kind of flow structure. Furthermore, a vast majority of academic papers would rather link the flow structure with various losses in the flow. In these works, qualitative description instead of quantitative relationship is overwhelming. The scale of the flow structure is another important subject discussed in fluids engineering. The scale is also an important factor considered in selecting the research tool. Here, the microscopic scale is beyond the scope of this book. We are just concerned with the scales that are meaningful to engineering applications.

Apart from the external environment, the fluid itself is complex as well. In practice, completely pure liquid or gas has rarely been found. Instead, the mixture of two or three phases is ubiquitous. In some occasions, even the same phase has constituents of different dynamics properties. For example, as the flow of pure water carries sand particles of 50 μm and 1 mm in diameter, the motion characteristics of these two kinds of solid particles are considerably different. Meanwhile, the influence of solid particles on water varies with the particle size.

The research in fluids engineering can be classified into three categories, namely basic research, applied research and product development. In this context, there is no definite demarcation between basic and applied research. The contents of this book fall into the category of applied research. Research methods and practical cases are two major aspects that the authors are intended to elucidate. The purpose is to replenish the knowledge of complex flows witnessed in engineering. In particularly, for some flows that the existing theoretical and experimental support is not sturdy. As the flows are fully understood, the improvement in the design of relevant machinery or system is facilitated.

1.2 Strategies for Treating Complex Flows

1.2.1 Fundamental Flow Features

A fluid is defined as a material that deforms continuously under the influence of shear stress. There are numerous ways of classification of fluid flows. In a very general way, fluid flows are divided into laminar and turbulent flows. The majority of flows found in engineering applications are turbulent flows. Although possessing such a high popularity, turbulence has no formalized definition until today. Here, three classical definitions of turbulence or turbulent flow are cited, they are:

(1) Turbulence is an irregular motion which in general makes its appearance in fluids, gaseous or liquid, when they flow past solid surfaces or even when neighbouring streams of the same fluid flow past or over one another (This definition is given by von Karman in 1937, he quoted this from a lecture made by G.I. Taylor in 1927 [1]).
(2) Turbulent fluid motion is an irregular condition of flow in which the various quantities show a random variation with time and space coordinates, so that statistically distinct average values can be discerned (Hinze [2]).

(3) Turbulence is a three-dimensional time-dependent motion in which vortex stretching causes velocity fluctuations to spread to all wavelengths between a minimum determined by viscous forces and a maximum determined by the boundary conditions of the flow. It is the usual state of fluid motion except at low Reynolds numbers (Bradshaw [3]).

Although a thorough understanding of turbulence has not been achieved to date, general features of turbulence can be described. Since this book is not a book devoted to the mechanism of turbulence, the readers are advised to search detailed explanations of turbulence in published literature.

Flows can be divided into internal and external flows. For the former, the walls bounding the flow are critical for determining the flow pattern. For the latter, flows passing a body are of special interest. Regarding some simple flows such as flows in a straight pipe, available theoretical foundation is sufficient and clear; therefore, the accordance between experiment results and theoretical results is conceivable. Nevertheless, with respect to rather complex internal flows in rotating machinery such as pumps and hydraulic turbines, both theoretical and experimental achievements are limited. In this case, the validation of assumptions is undoubtedly important.

1.2.2 Common Methods for Treating Flow Issues

The first step in solving a complex problem in fluids engineering is to analyze the environment of the case. Often, it is necessitated to trace the origin of the problem and establish a complete and specific description of the problem. Then seeking theoretical support or proof seems to be the most important but intricate step, particularly as the existing knowledge is inaccessible. After theoretical preparations are finished, a plan for addressing the problem will take shape, hopefully, as soon as possible. Then it is time to find tools to implement the plan. In this context, experimental and numerical tools are always expected. As a matter of fact, a huge amount of literature published recently in fluids engineering recorded experimental or numerical studies.

The analytical solution of flow-governing equations should be granted high priority. Since that most flows encountered in fluids engineering are in the turbulent regime, obtaining analytical solutions of the governing equations is not easy. A general strategy is simplification. The non-linear governing equations are simplified according to the characteristics of the problem. In this connection, some flow features might be intentionally weakened or even eliminated with the simplification. Therefore, to adopt such an approach, the physical essence of the flow problems considered should be sufficiently comprehended.

Regarding experimental techniques, expensive ones are not always advantageous. If you just want to know the averaged velocity of wind, then you do not need to use a laser Doppler velocimeter (LDV). In this context, a prudent consideration of what you are focusing on and what you should obtain with the research tools is crucial. Nevertheless, in fluids engineering, high-level measurement accuracy and resolution are helpful in most cases. As we measure the flow field inside the pump, the first instrument coming into mind is often particle image velocimetry (PIV). Traditional tools such as three-hole velocity probes cannot meet the requirements in today's view.

Computational fluid dynamics (CFD) has developed rapidly in recent 20 years. Certainly, this is related to the advancement of computer hardware. It should be noted that the contributions of numerical algorithms, turbulence models and multiphase models should never be omitted. There have been so many textbooks iterating the fundamental knowledge of CFD; therefore, the details of CFD are not presented here. Commercial CFD software is popular among researchers, particularly the graduate students. We should not criticize them for just simply inputting predefined parameters and setting boundary conditions and then waiting for the numerical results. Instead, we should encourage them to dig into the code and to find the reason why numerical results are dependent on the algorithms, turbulence models and the convergence criteria. Apart from CFD, the finite element method (FEM) has attracted lots of attention in fluids engineering. The most prominent advantage of FEM is it targets the solid components and extracts detailed information of the solid components such as stress and strain. Provided that the loads of fluid are properly exerted on the solid components, the FEM results are more valuable than CFD results for innovating the product or evaluating the operation stability of the fluid machine or the whole system.

In practical engineering, what we see are various machines and pipes rather than flows. Therefore, for the designer and maintainer, they are not concerned with distributions of velocity or turbulent kinetic energy. Instead, they are eager to know more about the machine itself or the operation performance of the machine. How does the flow affect the operation of the machine? How to build a connection between the flow and solid components of the machine? How to optimize the design of real products based on the flow parameter distributions? These questions are sometimes daunting for both researchers and engineers.

Problems encountered in fluids engineering provide a thrust to the technical advancement. In general, practical cases in fluids engineering involve a wide range of knowledge; thus, the treatment of these cases necessitates a systematic strategy. During the process of investigating the cases, the viewpoints will be expanded, various methods attempted and wisdom is concentrated. In this context, we cannot hide ourselves behind the fluid exclusively and we must strive to assimilate the knowledge of other disciplines. For instance, noise is emitted from an operating fan. To suppress the noise, one has to find the source of the noise and then to seek the route of noise emission. During this process, the knowledge of acoustics must be mastered. This is a case that justifies the importance of inter-discipline communication.

One cannot expect the benefits of research are acknowledged immediately after the conclusions are obtained. Some research results or conclusions can be used in the optimization of real product. However, a considerable amount of research achievement might be sealed for many years or just reflected in academic papers.

1.3 Characteristics of Flow Problems

1.3.1 Flow Quantities

Flow quantities are the premise of describing flow characteristics. Commonly, the flow quantity can be decomposed into averaged and fluctuation parts. A common viewpoint in engineering states that averaged flow quantities are much more important than fluctuation quantities. This is in most occasions acceptable since that a high percentage of total energy loss is associated with averaged flow quantities. Nevertheless, in some cases, the importance of fluctuations exceeds that of the averaged flow quantity. For instance, transient shock due to pressure fluctuations can lead to the damage to instruments mounted in the pipe flow system.

Vorticity, the divergence of velocity, is the flow quantity used extensively in fluids engineering. Vorticity distributions have been frequently used to express flow structures. For the optimal design of mechanical components wetted by fluid, the suppression of vorticity has often been taken as an objective. It is noteworthy that two-dimensional flows often serve as a simplification with respect to the original three-dimensional flows. Meanwhile, cross sections are frequently sliced from a three-dimensional flow field. In these two cases, the vorticity components should be used instead of the whole vector.

Turbulent fluctuations are important for describing the inherent feature of flow phenomena. To capture turbulent fluctuations, the temporal and spatial resolutions of the tool used must be very high. Alternatively, small temporal and spatial scales should be reached. The most popular turbulent quantities used today is the turbulent kinetic energy and the turbulent kinetic energy dissipation rate. In published reports, the two turbulence quantities have been used widely. Essentially, velocity distributions are the base of calculating vorticity or vorticity components; fluctuations of velocity are constituents of the turbulent kinetic energy. In this context, the proper orthogonal decomposition (POD) method has been adopted to extract the flow patterns that assume different percentages of total flow energy [4].

1.3.2 Unsteadiness

Flows can be divided into steady and unsteady flows. For steady flows, flow quantities at fixed position do not vary with time. Therefore, the time-related terms in the governing equations can be neglected. Meanwhile, for experimental treatment, in terms of both the instrument selection and experiment rig construction, convenience is appreciable. In contrast, unsteady flows are much more intricate. To obtain physically true flow information, the data-acquisition instrument should possess adequately high temporal resolution or alternatively, the response to the transient variation of flow quantities is quick enough. A typical case is the unsteady flows in the pump. Researchers attempted to measure pressure fluctuations in the pump with high-frequency pressure transducers. Then the data acquired were processed using the method of fast Fourier transformation (FFT).

1.3.3 Symmetry

Some flows are characterized by symmetry. For instance, as flow passes around a cylinder with a rather low Reynolds number, apparent symmetry is witnessed. In addition, the flow confined in a circular pipe is generally deemed as symmetrical relative to the axis of the pipe. This kind of flows are relatively easy to address. Nevertheless, the majority of flows in fluids engineering are non-symmetrical. In this case, corresponding theoretical support is not sound and even suffers from controversy.

1.3.4 Stability

For researchers, stability is an issue that considerably enhances the research difficulty. A very pertinent example is jet. As jet progresses, the surrounding fluid penetrates the jet stream consistently. During this process, the stability of the jet is undermined. Eventually, the jet cannot keep its original integrity. Thus far, the problem of flow stability is intricate. The reason of the stability loss is intricate; and the measures for capturing the initialization of instability are difficult to devise.

Solid boundary is extremely meaningful for large-scale flow structures. Meanwhile, loss of flow stability often stems from near-wall flows. From an academic view, near the wall, non-linear flow essence might be revealed and the linear relationship between the flow velocity and the distance from the wall seems ideal. In the flow passages inside the impeller pump, flows are affected significantly by the curved blade and passages. Meanwhile, the rotating blades impose another influence on flows inside the passage.

1.3.5 Symmetry, Intermittency and Periodicity

Some flows are characterized by symmetry, which is exemplified with the flow passing around a cylinder at a rather low Reynolds number. For these flows, the governing equations can be simplified. Meanwhile, for numerical simulation, the computation domain can be simplified as well. Nevertheless, in some cases, the seemingly symmetry is in fact not true. For instance, as a waterjet issued from a circular nozzle into ambient air. General viewpoint is that the jet flow is symmetrical with respect to the axis of the jet stream. Nevertheless, the disturbance from ambient air and the mixing of water and air at the jet edge might ruin the symmetry. Therefore, in an averaged manner, we may obtain symmetry flows; but for instantaneous flow parameter distributions, symmetry does not hold.

Intermittent flow phenomena are frequently witnessed in fluids engineering. We still use waterjet as an example. As waterjet is issued form a nozzle, the possibility of an intermittent jet is high. Some reports argue that the frequencies of the driven pump are the most influential factor underlying the intermittency. Some believe that the geometry of the flow passage in the nozzle plays a critical role. To treat this kind of flow phenomena, the concept of frequency is often introduced. Furthermore, not just an assessment of the intermittency, but also the relationship between the intermittency and flow parameters and the characteristic scale are anticipated to be explained.

As pump blades pass the volute tongue periodically, complex impeller-tongue interaction will be imposed on the flowing medium between the impeller and the volute tongue. This is just one simple case of periodic flows in fluids engineering. At present, the combination of rotating and stationary components in fluid machinery is diverse; therefore, the complexity of flow patterns in the narrow space between the neighbouring components is perceivable. In practice, for flows with periodic behavior, of interest are the flow structures and their excitations. This is also a topic of significance from both academic and engineering aspects. Most importantly, the excitation or periodic flows results in vibration and noise, sometimes, even the fatigue failure of mechanical components can be ascribed to the excitation.

1.3.6 Phase Change

When a substance transforms from one phase such as solid, liquid or gas to another phase, phase-change phenomenon occurs. In fluids engineering, phase change occurs occasionally. In this context, we will not go further in the chemical aspect of phase change. Meanwhile, in this book, the gas flow is not included and we will deal with phase-change phenomena in liquid flows.

Specific environment nurtures specific flow phenomenon. Cavitation is a phase-change phenomenon occurring in liquid. Cavitation is the formation of vapor cavities in a liquid as the static pressure is reduced to some threshold value in the liquid. When subjected to higher pressure, the cavities implode and might generate an intense shock wave, which can damage the adjacent objects. Cavitation can be detrimental to the performance of the hydraulic turbines and impeller pumps. Cavitation can undermine the thrust produced by a propeller. In some occasions, cavitation can severely damage a mechanical component and the operation of the whole unit must be halted. However, cavitation sometimes is helpful in disinfection and industrial cleaning. A pertinent example is the underwater cleaning using cavitating waterjet. Overall, the goal of studying cavitation is to control cavitation. In this book, cavitation will be analyzed in detail in different cases.

1.3.7 Flows in an Integrated System

Studies on flows can be limited to a single valve or pump, but practical engineering does not permit the isolation of flows from other components of the system. For instance, the flow system shown in Fig. 1.1 contains not just a single pump but also the valves, pipes and other auxiliary components. For such a system, only paying attention to flows in the pump will lead to non-physical results. Essentially, the operation of the pump is subjected to the resistance characteristics of the system. More specifically, the flow rate of the pump is not determined by the pump itself but by the whole system. Moreover, flow conditions at the inlet and outlet of the pump are affected by the system as well. Therefore, for numerical simulation, the assumption of uniform inflow and constant pressure at the pump outlet deviates from reality to some extent. To describe the flow characteristics of this system, the whole system should be modeled and analyzed.

Another example demonstrating the flow integrity is the prefabricated pump station (PPS), which is popular today because of its flexibility in installation and maintenance. The tank is a major component of PPS. In the tank, one, two or even three impeller pumps are mounted. Adjacent to the tank, inflow pipe and outflow pipe are equipped. Certainly, valves and filters are necessary for such a system.

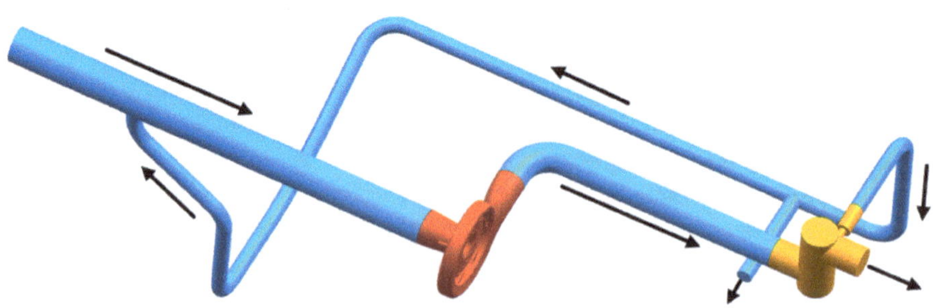

Fig. 1.1 The pipe flow system

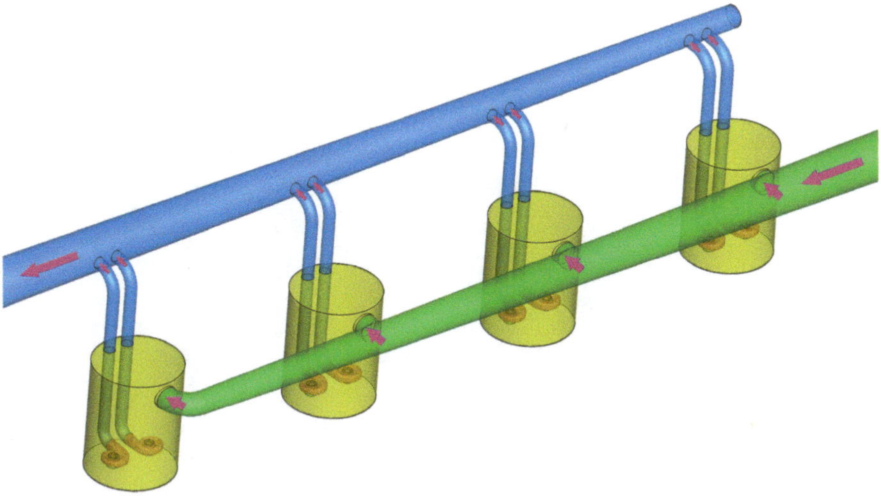

Fig. 1.2 Geometric model of a prefabricated pump station

Sometimes, to adapt to the transportation of medium of large flow rate, more than one tank is deployed. In Fig. 1.2, a schematic view of a prefabricated pump station with four tanks is shown. In each tank, there are two delivery pumps. For such a complex flow system, a comprehensive consideration of all pumps and tanks as well as the pipes is required. Actually, flow rate allocation among the four tanks influences significantly the operation of the pumps. Meanwhile, in each tank, the flow rates of the two pumps are possibly not equal. In addition, in each tank, the interaction between the inflows of the pumps and flows in the tank is related to both the pumps and the wall of the tank. These should be synthetically considered.

References

1. Pozrikidis C. Fluid dynamics, theory, computation, and numerical simulation. Berlin: Springer; 2017.
2. Hinze JO. Turbulence—an introduction to its mechanism and theory. New York: McGraw-Hill; 1959.
3. Bradshaw P. An introduction to turbulence and its measurement. Pergamon; 1971.
4. Berkooz G, Holmes P, Lumley JL. The proper orthogonal decomposition in the analysis of turbulent flows. Ann Rev Fluid Mech. 1993;25:539–575.

Chapter 2
A Brief Overview of Research Methods

Abstract Theoretical, experimental and numerical methods are three primary methods that are commonly used in fluids engineering. For the theoretical method, it requires a sound base of mathematical and mechanical knowledge. Meanwhile, the gap between theoretical results and applications is often remarkable. In contrast, the latter two methods can be easily exercised and the results can be transplanted into practical design. In this chapter, a brief overview of the two methods is presented. For each method, we do not intend to trace its origin or to explain its fundamental principles; these have been documented in detail. Only those contents that are much related to fluids engineering are presented here. In the following chapters, different cases will be introduced and the function of these methods will be substantiated then.

2.1 Introduction of Experiment

2.1.1 Definition of Experiment

Generally speaking, measurement is a process of gathering information from a physical world and comparing this information with agreed standards. Measurements are essential activities for observing and testing scientific and technological investigations.

Measurements are carried out using instruments, which are designed and manufactured to fulfill specific tasks. In this context, for the flow measurement, a focused subject in this book, conventional and advanced measurement techniques and instruments coexist. Pressure, velocity and temperature are three quantities of significance in incompressible flows. In this book, temperature is not the major concern; instead, pressure and velocity are emphasized. For velocity measurement, non-intrusive measurement techniques have exhibited distinct advantages relative to other techniques. Regarding pressure measurement, various dynamics sensor are used as the primary tools to respond to pressure fluctuations. Overall, the advance and development in measurements, instrumentation, and sensors have facilitated the deep

C. Kang et al., *Methods for Solving Complex Problems in Fluids Engineering*,
https://doi.org/10.1007/978-981-13-2649-3_2

understanding of complex flows in recent years. Moreover, the instruments can be connected together using wired, optical, or wireless networks. The development of the control software behind these networks plays an important role as well.

The task of the experiment in fluids engineering generally includes three aspects: monitoring flow quantities or flow patterns, control of flow data acquisition or image capturing, analysis of obtained results. For the first aspect, it refers to situations where the measurement apparatus is used to keep track of flow quantities or images. As we measure the wind speed using a small handheld anemometer, or we measure the water velocity in an open channel using a Pitot tube, the exercise falls into this category. The control of data acquisition or image capturing requires a systematic consideration of the setting of the experimental parameters. This is based on the specific cases considered. As for the analysis of the obtained results, both researchers and engineers need to be proficient in conducting this job. This stage is very important for solving an engineering problem or designing a more favorable product or system.

2.1.2 Experimental Instruments

It is perceivable that the measurement apparatus is crucial for the experimental work. In this book, the applications of measurement apparatuses are presented in a rather straightforward manner. The authors do not intend to reiterate fundamental principles underlying the apparatuses. Meanwhile, some advanced techniques such as magnetic resonance imaging (MRI) have been used in flow visualization, but these techniques have not been widely acknowledged; they will not appear in this book. Instead, commonly used instruments such as particle image velocimeter are used in this book to address flow issues and to disclose flow details.

Optical velocity measurement techniques are relatively advanced compared to the Pitot tube or other velocity probes. Currently, the laser Doppler anemometry (LDA) and particle image velocimetry (PIV) are two popular non-intrusive measurement techniques used in flow research. In some occasions, to identify small-scale flow structures, micro-PIV is put into use. As flow structures are fully three-dimensional, tomo-PIV can satisfy the requirement of flow structure construction. As the flow velocity is high, time-resolved PIV can yield results of sufficiently high temporal resolutions. With time-resolved PIV, the shooting frequency of the camera can be as high as 10,000 Hz. In contrast, ordinary PIV captures consecutive images at the frequencies from tens to more than 100 Hz.

The preparation of non-intrusive flow measurement or flow visualization experiments is not an easy task. First, the experiment model should be made transparent, or at least the incident light can penetrate the flow field and the reflected light can be received concurrently. In general, the experiment model can be made of plexiglass. Nevertheless, as the experiment model involves curved walls, the manufacturing methods and cost should be considered carefully. Sometimes the strength of the transparent model cannot meet the requirements of the flow load. In this case, the adoption of the transparent model is impractical. Additionally, as the incident light enters the flow field via curved transparent wall, the difference of the refractive index on the two sides of the curved wall will induce optical distortion. This is why we see the scenario where a circular transparent pipe is enclosed by a transparent rectangular pipe. Water is filled in the circular pipe as well as in the space between the two pipes. The purpose is to measure the flow velocity in the circular pipe.

As for turbulent fluctuation measurement, laser Doppler velocimetry is an unparalleled measurement tool. LDA uses Eulerian approach and sets a measurement volume in the flow field. Tracing particles passing through the measurement volume are recorded. As the velocity acquisition in one measurement volume is finished, the optical focus is shifted to another position in the flow field. Between PIV and LDA, similarity and difference are distinct. If you are intended to obtain the flow pattern or flow parameter distributions at the same moment, PIV is undoubtedly a better choice.

Processing and analyzing experimental results are an important step in experimental studies. We still take PIV as an example. Based on the consecutive images displaying the luminous tracing particles, an examination of the parameter setting should be performed to ensure that enough velocity vectors are covered in each inquiry window. Hence, averaged flow quantities should be calculated based on a sufficiently large group of transiently recorded images. Flow characteristics cannot be described without a comprehensive analysis of distributions of averaged and instantaneous flow quantities. In some cases, the dimensions of the monitored window of PIV are limited; therefore, an entire view of the flow cannot be obtained simultaneously. In this context, several images corresponding to different monitored windows might be stitched together to provide a full view of the flow.

For multiphase flows, the application of existing experimental techniques encounters many barriers. Multiphase flows encompass a diversity of flows such as the bubbly flow, the solid-liquid flow and the gas-liquid flow. For instance, the difficulty in dealing with the bubble flow is appreciable. For the observation of the flow patterns of the bubbly flow, high-speed photography is probably the most feasible approach that has ever been invented. The instantaneous pictures captured with short time intervals facilitate the seeking of unsteady flow mechanisms that otherwise cannot be illustrated. Often, many researchers are daunted by the huge number of the images recorded. In this context, effective image processing requires a powerful tool. The majority of the software packages affiliated to the high-speed camera cannot meet the requirements of image processing. In this context, the development of a specific code is necessitated. Such a code not just can extract statistical information from the images but also can identify distinct elements in the image. For bubble images, even the profiles of individual bubbles can be recognized using the developed code.

Apart from flow velocity measurement, static pressure is also a flow quantity of significance. Nevertheless, available optical instruments cannot measure pressure distribution in the flow field. Instead, pressure probes and transducers are used overwhelmingly at present to realize pressure measurement. Since the disturbance of the pressure probe to the flow is inevitable, it is expected to minimize the size of the probe. Another way of measuring pressure is through deploying pressure transducers in the wall bounding the fluid flow. Such a scheme has been adopted widely in fluid machinery to measure near-wall pressure fluctuations. Certainly, commercial instruments of pressure measurement are available. For them, the competition resides in instrument parameters such as the maximum data-acquisition frequency, the spatial resolution and the number of data channels.

2.1.3 *Fidelity of Experiment*

Reliable experimental results should be repeatable and reproducible. This means that the experimental results are not incidentally obtained; furthermore, with another experiment rig, which is identical with the one yielding the experimental results, same results should be obtained. For fluid flow experiments, particularly the experiments of liquid flows, high repeatability is difficult to achieve in view of the essence of turbulence. Therefore, the experimental rig, including all the instruments, should be checked carefully prior to the experiment. Meanwhile, the potential factors that impose a negative effect on the experiment should be avoided. For instance, the pulsation in the power grid will undermine the stability of the rotational speed of the motor. Therefore, the experiment should not be conducted in the peak period of electricity consumption. Actually, uncertainties are unavoidable for experiment. In flow experiments, uncertainties within some limit are acceptable. The exact limit depends on relevant experiment standards.

It is generally acknowledged that in most cases, experimental results are more convincible that numerical ones. Nevertheless, the cost of experiment is high, which can be immediately justified from the prices of the optical measurement instruments and the assistant device such as the traverse system. Moreover, the period of experiment preparation is fairly long. Therefore, an elaborate design of the experiment scheme and corresponding preliminary validation are important. Furthermore, as the experiment is launched, there are still some unpredictable factors that will impede the progress of the experiment.

2.2 Numerical Methods for Flows

At present, no engineering field can take evasive action against the computer technology. The task of scientific computation is undertaken by computers. A huge amount of data are being processed everyday by computers. In fluids engineering, the introduction of computers enable the implementation of numerical simulation, which involves flow simulation and structural simulation. For the former, it is performed using the computational fluid dynamics technique. For the latter, the finite element method (FEM) is prevalent. Thus far, more than a dozen of commercial CFD codes are being used in fluids engineering, and there are other in-house codes that assume the same role. Although some researchers focusing on experiments do not believe numerical results at all, the pace of the development of CFD has never been halted any more.

Computational fluid dynamics is the analysis of systems involving fluid flow, heat transfer and associated phenomena such as chemical reactions by means of computer-based simulation [1]. The early development of CFD in the 1960s and 1970s was driven by the needs of aerospace community. Then CFD techniques progress quickly. Currently, CFD has penetrated all disciplines where the fluid flow plays an important role. Meanwhile, CFD has become a research tool as well as a design tool in fluids engineering [2]. CFD results are analogous to the results obtained in a laboratory. Nevertheless, the water tunnel is a heavy device and cannot be carried around with ease. In comparison, the computer code is easily accessible. Even as the researcher has no code in hand, he can operate the code via internet.

2.2.1 Governing Equations

No matter how complex the flow is, it follows some rules. The conservation laws of mass, momentum and energy are the most fundamental base of fluid flow. Therefore, the most important task of a CFD code is to obtain the numerical solution of the equations associated with the three conservation laws. Meanwhile, distributions of flow quantities in defined computational domain can be obtained using CFD. As mentioned before, the flow quantity is composed of an averaged part and a fluctuation part. Here, the flow quantity ϕ is time dependent and can be thought of as the sum of steady mean component Φ and a time-varying fluctuating component ϕ'. So we write:

$$\phi(t) = \Phi + \phi'(t) \tag{2.1}$$

The time average of the fluctuations ϕ' is, by definition, zero:

$$\overline{\phi'} = \frac{1}{\Delta t}\int_{0}^{\Delta t} \phi'(t)\mathrm{d}t = 0 \tag{2.2}$$

Hence, the mean Φ of a flow quantity ϕ is given as follows:

$$\Phi = \frac{1}{\Delta t}\sum_{0}^{\Delta t} \phi(t)\mathrm{d}t \tag{2.3}$$

Theoretically, we should take the limit of time interval Δt approaching infinity. Nevertheless, Δt is sufficiently large as it exceeds the time scales of the slowest variations of ϕ, which are associated with large-scale flow structures. This formula of the mean of a flow quantity is adequate for steady mean flows. In time-dependent flows, the mean of a quantity at time t is taken to be the average of the instantaneous values of the quantity over a large number of repeated identical experiments; this procedure is called ensemble average. Here, we do not consider the change of the fluid compressibility and the internal energy. Therefore, the variations in density and temperature are neglected.

For flow velocity vector $\boldsymbol{u}$, we can express it in the following form:

$$\boldsymbol{u} = \boldsymbol{U} + \boldsymbol{u}' \tag{2.4}$$

Since flow velocity components fluctuate drastically in some turbulent flows, we often use root-mean-square (rms) flow velocity components:

$$\phi_{\mathrm{rms}} = \sqrt{\overline{(\phi')^2}} = \left[\frac{1}{\Delta t}\int_0^{\Delta t} (\phi')^2 \mathrm{d}t\right]^{0.5} \tag{2.5}$$

The rms values of the velocity components can be measured with experimental techniques such as laser Doppler velocimetry or hot-wire anemometry. The turbulent kinetic energy (per unit mass) is defined as:

$$k = \frac{1}{2}\left(\overline{u'^2} + \overline{v'^2} + \overline{w'^2}\right) \tag{2.6}$$

The turbulent intensity is related to the turbulent kinetic energy and a reference mean flow velocity:

$$T_i = \frac{\left(\frac{2}{3}k\right)^{0.5}}{U_{\mathrm{ref}}} \tag{2.7}$$

Instantaneous continuity and Navier-Stokes equations for an incompressible flow with constant viscosity are given by:

$$\mathrm{div} u = 0 \tag{2.8}$$

$$\begin{aligned}\frac{\partial u}{\partial t}+\operatorname{div}(u\boldsymbol{u}) &= -\frac{1}{\rho}\frac{\partial p}{\partial x}+\nu \operatorname{div} \operatorname{grad} u\\ \frac{\partial v}{\partial t}+\operatorname{div}(v\boldsymbol{u}) &= -\frac{1}{\rho}\frac{\partial p}{\partial y}+\nu \operatorname{div} \operatorname{grad} v\\ \frac{\partial w}{\partial t}+\operatorname{div}(w\boldsymbol{u}) &= -\frac{1}{\rho}\frac{\partial p}{\partial z}+\nu \operatorname{div} \operatorname{grad} w\end{aligned} \tag{2.9}$$

where Cartesian co-ordinates are used and the velocity vector has x, y and z components.

To study fluctuations, we use the sum of a mean value and a fluctuating component to replace the flow quantities in Eq. (2.9).

After time-average processing, time-averaged momentum equations are obtained:

$$\begin{aligned}\frac{\partial U}{\partial t}+\operatorname{div}(U\boldsymbol{U}) &= -\frac{1}{\rho}\frac{\partial p}{\partial x}+\nu\operatorname{div}(\operatorname{grad}U)+\left[-\frac{\partial\overline{u'^2}}{\partial x}-\frac{\partial\overline{u'v'}}{\partial y}-\frac{\partial\overline{u'w'}}{\partial z}\right]\\ \frac{\partial V}{\partial t}+\operatorname{div}(V\boldsymbol{U}) &= -\frac{1}{\rho}\frac{\partial p}{\partial y}+\nu\operatorname{div}(\operatorname{grad}V)+\left[-\frac{\partial\overline{u'v'}}{\partial x}-\frac{\partial\overline{v'^2}}{\partial y}-\frac{\partial\overline{v'w'}}{\partial z}\right]\\ \frac{\partial W}{\partial t}+\operatorname{div}(W\boldsymbol{U}) &= -\frac{1}{\rho}\frac{\partial p}{\partial z}+\nu\operatorname{div}(\operatorname{grad}W)+\left[-\frac{\partial\overline{u'w'}}{\partial x}-\frac{\partial\overline{v'w'}}{\partial y}-\frac{\partial\overline{w'^2}}{\partial z}\right]\end{aligned} \tag{2.10}$$

This equation set is called the Reynolds equations. The turbulent stresses are termed Reynolds stresses, among which three normal stresses, $-\rho\overline{u'^2}$, $-\rho\overline{v'^2}$ and $-\rho\overline{w'^2}$, are non-zero, the shear stresses $-\rho\overline{u'v'}$, $-\rho\overline{u'w'}$ and $-\rho\overline{v'w'}$ are associated with correlations between different velocity components. If u' and v' are statistically independent, $\overline{u'v'}$ would be zero. The shear stresses are non-zero and are usually much larger than viscous stresses in a turbulent flow. It is the task of turbulence modelling to develop computational procedures of sufficient accuracy and generality for engineers to predict the Reynolds stresses.

For most engineering issues, it is unnecessary to resolve the details of turbulent fluctuations. Only the effects of the turbulence on the mean flow are usually considered.

A turbulence model is a computational procedure to arrive at the closure of the system of the flow-governing equations. For a turbulence model, wide applicability, high accuracy, simple and economical to run are primary traits.

Classical models are based on time-average Reynolds equations. The typical models in this category are zero-equation models, two-equation modes, Reynolds stress equation models and algebraic stress models. Large eddy simulation can be deemed as a turbulence model where the time-dependent flow equations are solved for the mean flow and the large eddies. The effects of the smaller eddies are modelled. It is perceivable that the large eddies interact with the mean flow and contain most of the energy so large eddy simulation is a good model. However, at present, large eddy simulation has only been used under simple boundary conditions.

2.2.2 Turbulence Model

Turbulent model is necessary for attaining the closure of flow-governing equations. Distinct advantages of RANS-based turbulent models in the simulation of engineering flows have been acknowledged extensively [3]. Thus far, zero-equation, one-equation and two-equation turbulence models are prevalent in fluids engineering. Nevertheless, there is no turbulence model that has been proved to be superior to others in the treatment of various flows. Some turbulence models are advantageous for flows rotating with the impeller, and some particularly adapts to the jet flow issuing from a small nozzle.

The advantages of adaptability of the turbulence model maybe hold only for certain cases. For instance, the two-equation shear stress transport (SST) k–ω turbulence model is often used in the treatment of flows in impeller machinery [4]. The SST k–ω turbulence model combines the merits of k–ε and k–ω models and can handle both free stream flow and the flow near the impeller blade with high reliability. Furthermore, the SST k–ω turbulence model is particularly suitable to predict flow phenomena dominated by curved solid walls and rotating frame [5]. With respect to the simulation of flows in the impeller pump, convincible results have been obtained with the SST k–ω turbulence model [6].

The renormalization group (RNG) k–ε turbulence model, firstly proposed by Yakhot and Orszag in 1986, represents an improvement relative to the standard k–ε turbulence model [7]. This model is established based upon fuzzy mathematics principles, and the parameters incorporated in this model are deduced from related formulae instead of empiricism or experiments. Along this line, the resultant equation of turbulent kinetic energy dissipation rate ε differs from its counterpart in the standard k–ε turbulence model. The transport equations of turbulent kinetic energy k and ε are:

$$\rho\frac{Dk}{Dt} = \frac{\partial}{\partial x_i}\left[\left(\mu + \frac{\mu_t}{\sigma_k}\right)\frac{\partial k}{\partial x_i}\right] + G_k - \rho\varepsilon \tag{2.11}$$

$$\rho\frac{D\varepsilon}{Dt} = \frac{\partial}{\partial x_i}\left[\left(\mu + \frac{\mu_t}{\sigma_\varepsilon}\right)\frac{\partial \varepsilon}{\partial x_i}\right] + C_{\varepsilon 1}\frac{\varepsilon}{k}G_k - C^*_{\varepsilon 2}\rho\frac{\varepsilon^2}{k} \tag{2.12}$$

and the variable $C_{\varepsilon 2}^{*}$ in Eq. (2.12) is defined as:

$$C_{\varepsilon 2}^{*} = C_{\varepsilon 2} + \frac{C_{\mu}\rho\eta^{3}\left(1 - \frac{\eta}{\eta_{0}}\right)}{1 + 0.012\eta^{3}} \tag{2.13}$$

where dimensionless strain-rate parameter η is obtained from $\eta = Sk/\varepsilon$, where S denotes mean strain rate.

Additionally, μ and μ_t are viscosity and turbulent viscosity, respectively. G_k is a turbulent kinetic energy production term related to viscous force.

Predefined values of the constants in Eqs. (2.11) to (2.13) are:

$$C_{\mu} = 0.0845,\ C_{\varepsilon 1} = 1.42,\ C_{\varepsilon 2} = 1.68,\ \sigma_{k} = 0.72,\ \sigma_{\varepsilon} = 0.75,\ \eta_{0} = 4.38$$

In flow regions with low strain rates, $\eta < \eta_0$, the eddy viscosity calculated with the RNG k–ε model is higher than that determined using the standard k–ε model, and this trend is reversed in high-strain-rate flow regions characterized by $\eta > \eta_0$.

2.2.3 *Turbulent Characteristics*

Free turbulent flows such as jets, wakes and mixing layers are often witnessed in fluid engineering. These are generally deemed as the simplest types of engineering turbulent flows. Velocity changes rapidly across a thin layer, then transition to turbulence occurs after a short distance from the nozzle. At the interface between high-velocity jet stream and the surrounding fluid, turbulence causes vigorous mixing of adjacent fluid layers and rapid widening of the region across which velocity changes take place.

Compared to free turbulent flows, turbulent flows near solid walls exhibit complex flow behavior and turbulent structures. In fluids engineering, these flows are more common. Regarding flows along solid boundaries, there is usually a substantial region of inertia-dominated flow far away from the wall and a thin layer within which viscous effects are predominant. In impeller machinery with narrow and curved blade passages, the inertia-dominated region is difficult to define.

Close to the wall, the flow is influenced by viscous effects and does not depend on free stream parameters. The mean flow velocity only depends on the distance y from the wall, fluid density ρ and dynamics viscosity μ and wall shear stress τ_{w}. Thus:

$$U = f(y, \rho, \mu, \tau_{\mathrm{w}}) \tag{2.14}$$

Dimensional analysis yields:

$$u^{+} = \frac{U}{u_{\tau}} = f\left(\frac{\rho u_{\tau} y}{\mu}\right) = f(y^{+}) \tag{2.15}$$

This is the law of the wall and includes two dimensionless groups u^+ and y^+. The friction velocity is given by

$$u_{\tau} = \left(\frac{\tau_{\mathrm{w}}}{\rho}\right)^{0.5} \tag{2.16}$$

The fluid closest to the wall is dominated by viscous shear. In such a thin layer, y^+ is expected to be less than 5 and turbulent shear stress is absent. The shear stress in this layer is approximately constant and equal to the wall shear stress. Thus

$$\tau(y) = \mu \frac{\partial U}{\partial y} \cong \tau_{\mathrm{w}} \tag{2.17}$$

After integration with respect to y and application of boundary condition $U = 0$ if $y = 0$, we obtain a linear relationship between the mean velocity and the distance from the wall:

$$U = \frac{\tau_{\mathrm{w}} y}{\mu} \tag{2.18}$$

This is equivalent to:

$$u^{+} = y^{+} \tag{2.19}$$

Outside the viscous sublayer ($30 < y^+ < 500$), a region exists where viscous and turbulent effects are of equivalent importance. The shear stress varies slowly with the distance from the wall. Within this inner region, it is assumed that the shear stress is constant and equal to the wall shear stress. As the von Karman constant κ is introduced, the relationship between u^+ and y^+ in this layer is expressed as:

$$u^{+} = \frac{1}{\kappa} \ln y^{+} + B = \frac{1}{\kappa} \ln(Ey^{+}) \tag{2.20}$$

This is the log-law formula.

In the outer layer, the following formula is given:

$$\frac{U_{\max} - U}{u_\tau} = \frac{1}{\kappa} \ln\left(\frac{y}{\delta}\right) + A \tag{2.21}$$

where δ is the boundary layer thickness; A is a constant. This formula is called the law of the wake. This formula is true as $y/\delta > 0.2$.

2.2.4 Cavitation Model

Some phenomena in the flow field need to be modeled so the code can be guided to reveal the occurrence, development and attenuation of the phenomena. A typical phenomenon is cavitation. Cavitation modeling has remained a subject of active research in recent years [8].

At present, two aspects are of particular interest in the treatment of cavitation related issues. One is the tracing of the interface between cavities or cavitation bubbles and ambient liquid [9]. Following this route, details of cavity geometry are expected be revealed. Both experimental and numerical efforts have been consumed considerably in this aspect. Nevertheless, the investigation in this aspect is seemingly more promising for single bubble or sparsely distributed bubbles. With respect to cavitation cloud or inter-connected cavities, the task of discriminating phase interface is really daunting. The other aspect is the pursuit of the relationship between the underlying liquid flow and cavitation. It falls into the category of macroscopic research. Local flow patterns determine the cavity topology. Cavities, in turn, impose a perturbation on the local flow. A pertinent event is the cavitation phenomenon occurring in centrifugal pumps [10]. Turbulent flows in the impeller pump are subjected to the influence of blade rotation and curved solid walls, so flow structures with various spatial and temporal scales are desirable [11]. Cavitation surviving in such a circumstance is impressed more or less by the intricate liquid flow, as enhances the difficulty in depicting the cavitation feature of the centrifugal pump.

The numerical model of cavitation plays a crucial role in this connection, but no consensus about the most reliable cavitation model has been attained. The importance of those factors underlying cavitation such as compressibility and medium property still suffers from dispute [12]. Meanwhile, the progress of cavitation model has never halted, and even the model of cavitation erosion has been developed and put into practice [13].

Until now, unanimous opinion on the selection of cavitation model has not been reached, although the comparison between different cavitation models has been performed frequently [14–16]. Furthermore, the cavitation model should well match the turbulent model to ensure the effect of local flow on cavitation is properly treated [17]. This can be explained with two cases, one is cavitation near the propeller blades and the other is cavitation on the suction surface of a hydrofoil. Apparently, the carrier flow propelled by the rotating blades differs considerably from the flow around a stationary hydrofoil [18]. Consequently, the cavitation patterns are different, which demonstrates the influential effect of the flow on cavitation. Certainly, the physical essence of cavitation itself deserves a special consideration as well [19].

Here, the cavitation model of Zwart-Gerber-Belamri (ZGB model) is explained as a representative [20]. This model proves to be feasible under various cavitation conditions and has been embedded in the commercial CFD code of ANSYS CFX. Most importantly, the ZGB model is fully compatible with the RANS-based simulation, as facilitates the establishment of the relationship between flow structures and their influence on cavitation in the condensate pump.

With ZGB model, the mass transfer between liquid and vapor phases takes primarily the forms of evaporation and condensation, which are governed by the transport equation based on liquid volume fraction α_1:

$$\frac{\partial}{\partial t}(\alpha_1 \rho_1) + \nabla \cdot \left(\alpha_1 \rho_1 \vec{u}_m\right) = \dot{m} \tag{2.22}$$

Condensation and evaporation in phase change and mass transfer are substantiated with [21]:

if $p < p_{\mathrm{v}}$,

$$\dot{m} = -C_{\mathrm{v}} \frac{3\alpha_n (1 - \alpha_2)\rho_2}{R_B} \sqrt{\frac{2}{3}\left(\frac{p_{\mathrm{v}} - p}{\rho_1}\right)} \tag{2.23}$$

and, if $p > p_{\mathrm{v}}$,

$$\dot{m} = C_{\mathrm{c}} \frac{3\alpha_2 \rho_2}{R_B} \sqrt{\frac{2}{3}\left(\frac{p - p_{\mathrm{v}}}{\rho_1}\right)} \tag{2.24}$$

In Eqs. (2.23) and (2.24), C_v and C_c are empirical constants for evaporation and condensation, and are often set equal to 50 and 0.01, respectively. R_B is the cavitation bubble radius and is generally set to 10^{-3} mm, α_n denotes the nucleation site volume fraction and is generally set to 5×10^{-4}. It is noteworthy that these settings have a significant effect on the numerical results obtained.

2.2.5 Other Issues

Grid generation is another critical step for the implementation of numerical simulation. This aspect is not detailed here since the knowledge of grid generation can be easily found in textbooks or on the internet. It is noteworthy that the numerical results can be considerably different as different grid schemes are adopted for the same case. A grid-independence examination is required. Meanwhile, the grid quality should be examined as the grid generation is finished. Sometimes, an excessive refinement of the grids near solid walls results in a huge and unacceptable grid number.

Some numerical tactics were developed to adapt to specific cases. For instance, to simulate flows driven by the rotating impeller, the method of moving reference frame (MRF) was invented. This method assumes that an assigned volume has a constant speed of rotation and the non-wall boundaries are surfaces of revolution. Therefore, the volumes between the impeller blades are designated as rotating with predefined rotational speeds. Another method, namely moving mesh method, allows for the simulation of the motion of blades which interact with the medium in the blade passages. Such a method relates the motion of the rigid blades with the deformation of the computational grids. These methods considerably facilitate the simulation of the flow exposed to rotating impeller. Moreover, the extension of the application of these methods has been widely acknowledged.

2.3 Some Limitations of CFD

Although CFD has gained enormous popularity in fluids engineering, its limitations are apparent. For instance, current CFD cannot properly handle the liquid flow that carries swarms of gas bubbles; it cannot capture the motion of each bubble and quantify the influence of bubble wake on adjacent bubbles. Another case that CFD cannot treat is the liquid metal flow that is consistently merged by solid particles, which come from the removal of the wall bounding the flow. In this case, CFD cannot predict the erosion of the wall by the liquid metal flow. The development of the models that can describe these intricate physical phenomena is sorely expected.

Most users of CFD techniques are just users instead of developers. Although some CFD code provides users an access to adjust the parameter in the numerical settings, most users do not know the core part of the code. Moreover, we hope the accuracy of CFD can be as high as possible, but this highly depends on the users' experience and strategies. Thanks to the widespread availability of open-source code such as Openfoam, researchers have the opportunity of supplementing or improving codes for specific flow problems. Such a laborious process often daunts new code developers.

From the academic viewpoint, the temporal and spatial accuracy in the simulation of turbulent flows has never reached its peak value. It has been acknowledged that turbulent flows are characterized by eddy motions of a wide range of length scales. If the fastest events take place with a frequency of the order of 10 kHz, we would need to discretize time into steps of about 100 μs. This is difficult to attain in most laboratories. Regarding engineering applications, the prediction of each eddy in the flow is not necessary. Instead, time-averaged properties such as mean velocities and mean pressures are desirable. This can be achieved with common

CFD strategies. In addition, from the viewpoint of engineering, the accuracy of CFD results is not expected to be very high. This is related to the accuracy of the instruments; highly accurate CFD results cannot find their correspondence in experimentally obtained data. For instance, the pressure measured in engineering is the averaged pressure over a small area of a diaphragm. In contrast, the pressure obtained with CFD is often the pressure at some certain point, which is obtained by mathematical strategies such as interpolation. Apparently, these two pressures cannot well match each other.

Experiment boosts the progression of CFD. Some CFD practitioners firmly believe that the codes they used and the numerical settings are reliable. In this connection, it is necessary to refer to some published benchmark experimental results to validate the numerical approach. In many occasions, there is no such result. Then some practitioners would carry out an experiment to validate the numerical scheme. In fluids engineering, the combination of numerical simulation and experiment is highly desirable. With numerical simulation, preliminary design can be conveniently and efficiently accomplished. With experiment, the performance of the designed product or system is witnessed, and the numerical scheme is validated. Hence, the optimization task can be fulfilled using numerical simulation. All in all, experiment should be performed whenever it is necessary.

2.4 Structural Analysis

The fluid flow is restricted by mechanical parts. CFD can solve flow quantities but cannot evaluate the safety and stability of the solid parts. For engineers, they are more concerned with the solid parts that with the flow. Therefore, considering only a single factor cannot seize the physical essence and multiple fields should be taken into account. A more complex topic is the fluid-structure interaction, which requires the consideration of the deformation of solid walls due to hydraulic loads as well as the change in the flow characteristics due to the variation in restrictive boundaries. Sometimes, the fluid-structure interaction is not significant. For instance, regarding centrifugal pumps of low specific speed, the coupling between flow and the pump impeller is often neglected. Nevertheless, for the axial-flow pump and mixed-flow pump with high specific speed, the blade deformation and consequent variation in the near-wall flow deserve much attention.

Derived from Newton's second law, the conservation equations of the solid part are expressed as:

$$\rho_s \ddot{d}_s = \nabla \cdot \sigma_s + f_s \tag{2.25}$$

where ρ_s denotes the density of structure, σ_s is the Cauchy stress tensor, f_s is the volume force vector, and $\ddot{d}_s$ is the acceleration vector.

Modal analysis is an effective method of evaluating dynamic characteristics of a structure [22]. Modal analysis also serves as a support for the analysis of structural vibration, vibration fault diagnosis and prediction.

According to Newton's second law and vibration theory, the modal equations of structural systems with multiple degrees of freedom are as follows:

$$[M]\{\ddot{u}\} + [C]\{\dot{u}\} + [K]\{u\} = \{F(t)\} \tag{2.26}$$

where $[M]$ is the mass matrix, $[C]$ is the damping matrix, $[K]$ is the stiffness matrix, $\{\ddot{u}\}$ is the acceleration vector, $\{\dot{u}\}$ is the velocity vector, $\{u\}$ is the displacement vector and $\{F(t)\}$ is the force vector.

Provided that the external excitation force is not considered, and the effect of structure damping is ignored, Eq. (2.26) can be simplified as:

$$[M]\{\ddot{u}\} + [K]\{u\} = 0 \tag{2.27}$$

The solution of Eq. (2.27) takes the form:

$$\{u\} = \{\emptyset\}e^{i\omega t} \tag{2.28}$$

A combination of Eqs. (2.27) and (2.28) yields the characteristic equation of the free mode of the structure:

$$[K]\{\emptyset\} = \lambda[M]\{\emptyset\} \tag{2.29}$$

where $\lambda = \omega^2$.

References

1. Versteeg H, Malalasekera W. An Introduction to Computational Fluid Dynamics. London: Pearson Education Limited; 1995.
2. Anderson JD. Computational Fluid Dynamics, The Basics with Applications. New York: McGraw-Hill; 1995.
3. Mani KV, Cervone A, Hickey JP. Turbulence modeling of cavitating flows in liquid rocket turbopumps. J Fluids Eng Trans ASME. 2017;139: 011301–1–011301–10.
4. Menter F. Two-equation eddy-viscosity turbulence models for engineering applications. AIAA J. 1994;32:1598–1605.
5. Jiang W, Li G, Liu P, Lei F. Numerical investigation of influence of the clocking effect on the unsteady pressure fluctuations and radial forces in the centrifugal pump with vaned diffuser. Int Commun Heat Mass Transfer. 2016;71:164–171.
6. Limbach P, Skoda R. Numerical and experimental analysis of cavitating flow in a low specific speed centrifugal pump with different surface roughness. J Fluids Eng Trans ASME. 2017;139:101201–1–101201–8.
7. Yakhot V, Orszag SA. Renormalization group analysis of turbulence. J Sci Comput. 1986;1:3–51.

8. Rodio MG, Abgrall R. An innovative phase transition modeling for reproducing cavitation through a five-equation model and theoretical generalization to six and seven-equation models. Int J Heat Mass Transf. 2015;89:1386–1401.
9. Roohi E, Zahiri AP, Zahiri AP, Passandideh-Fard M. Numerical simulation of cavitation around a two-dimensional hydrofoil using VOF method and LES turbulence model. Appl Math Model. 2013;37:6469–6488.
10. Gülich TF. Centrifugal pumps. 3rd ed. Berlin Heidelberg: Springer-Verlag; 2014.
11. Li X, Zhu Z, Li Y, Chen X. Experimental and numerical investigations of head-flow curve instability of a single-stage centrifugal pump with volute casing. Proc Inst Mech Eng Part A: J Power Energy. 2016;230:633–647.
12. Peng G, Okada K, Yang C, Oguma Y, Shimizu S. Numerical simulation of unsteady cavitation in a high-speed water jet. Int J Fluid Mach Syst. 2016;9:66–74.
13. Peters A, Sagar H, Lantermann U, el Moctar O. Numerical modelling and prediction of cavitation erosion. Wear. 2015;338–339:189–201.
14. Kozubková M, Rautová J, Bojko M. Mathematical model of cavitation and modelling of fluid flow in cone. Procedia Engineering. 2012;39:9–18.
15. Morgut M, Nobile E, Biluš I. Comparison of mass transfer models for the numerical prediction of sheet cavitation around a hydrofoil. Int J Multiph Flow. 2011;37:620–626.
16. Saha K, Li X. Assessment of cavitation models for flows in diesel injectors with single–and two–fluid approaches. J Eng Gas Turbines Power Trans ASME. 2016; 138:011504–1–011504–11.
17. Charrière B, Decaix J, Goncalvès E. A comparative study of cavitation models in a Venturi flow. Eur J Mech B/Fluids. 2015;49:287–297.
18. Kim J, Lee JS. Numerical study of cloud cavitation effects on hydrophobic hydrofoils. Int J Heat Mass Transf. 2015;83:591–603.
19. Zuo Z, Liu S, Liu D, Qin D, Wu Y. Numerical analyses of pressure fluctuations induced by interblade vortices in a model Francis turbine. J Hydrodyn. 2017;27:513–521.
20. Zwart PJ, Gerber AG, Belamri T. A two-phase flow model for predicting cavitation dynamics. In: ICMF 2004 International conference on multiphase flow, Yokohama, Japan; 2004.
21. Liu D. The numerical simulation of propeller sheet cavitation with a new cavitation model. Procedia Eng. 2015;126:310–314.
22. Liu M, Xia H, Sun Lin, Li B, Yang Y. Vibration signal analysis of main coolant pump flywheel based on Hilbert-Huang transform. Nucl Eng Technol. 2015;47:219–225.

Chapter 3
Submerged Waterjet

Abstract Successful applications of waterjet have been observed in engineering fields. For submerged waterjet, the interaction between the jet stream and ambient water is intricate. As the jet pressure is high, cavitation might occur; meanwhile, due to the resistance of surrounding water, the integrity of the jet stream will be ruined rapidly after the waterjet is issued from the nozzle. In this chapter, submerged waterjets driven at various pressures are discussed. Flow velocity is measured using particle image velocimetry technique; therefore, overall energy dissipation with the progression of the waterjet is evaluated. Flow patterns are constructed and analyzed. Pressure fluctuations in surrounding water excited by the waterjet are measured and explained. Meanwhile, the cavitation phenomenon arising in such a special environment is discussed. Not just cavity topology but also the cavitation erosion effects on the specimen impacted by the submerged waterjet are investigated. A comprehensive study of the submerged waterjet is presented.

3.1 Fundamental Features of Submerged Waterjet

The submerged waterjet is produced as the waterjet stream is injected into initially stationary water. The advantages of such a waterjet type are apparent and have been substantiated in applications such as underwater cleaning and sterilization [1]. It has been reported that the concept of the submerged jet is even extended to the oil leak during deep water drilling [2].

As jet pressure is low, the resistance of ambient water will hinder the jet development and even destroy the integrity of the jet stream. On the contrary, at high jet pressure, the waterjet stream can sustain for a long downstream distance away from the nozzle. Meanwhile, high-speed submerged waterjet provides a favorable environment for cavitation occurrence [3]. Low static pressure resides in the vortices at the interface between high-velocity waterjet stream and nearly stationary ambient water. In some cases, the submerged waterjet is used to produce cavitation bubbles which are then used to clean specimen surface [4].

C. Kang et al., *Methods for Solving Complex Problems in Fluids Engineering*,
https://doi.org/10.1007/978-981-13-2649-3_3

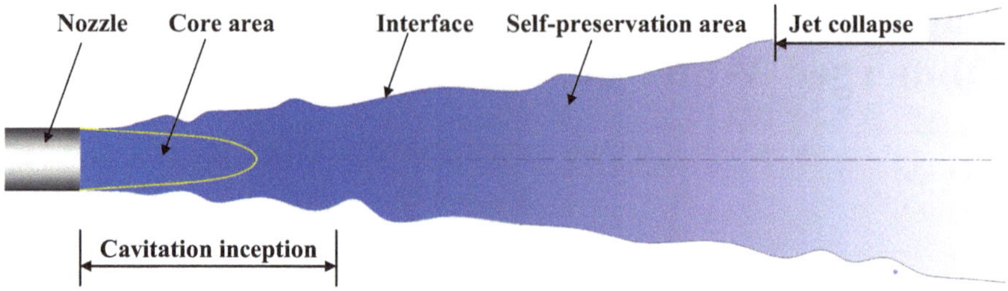

Fig. 3.1 Schematic view of a submerged waterjet streaming rightward

A schematic view of the submerged waterjet is shown in Fig. 3.1. The waterjet progresses from left to right. The intensive interaction between the jet stream and ambient water results in the wavy jet stream edge. Cavitation might occur near the nozzle, where the velocity difference or shear effect between the jet stream and surrounding water is remarkable, and then cavitation bubbles will be carried by the bulk flow. The core area of the jet stream is featured by highly concentrated kinetic energy, so the length of the core area increases with the jet pressure. Although the submerged waterjet will collapse eventually, a self-preservation stage with similar cross-sectional velocity distributions exists. In addition, the impingement effect of the submerged waterjet is of significance as the submerged waterjet serves as an effective surface hardening tool [5].

The visualization of the submerged waterjet is possible with optical techniques [6]. With high speed photography (HSP) introduced in Chap. 2, transiently varied phenomena in the waterjet are to be disclosed. Furthermore, a joint application of micro-lens and HSP enables a deep understanding of spatial and temporal characteristics of the tiny waterjet stream [7]. Regarding transient flow measurement, time-resolved PIV is highly advised. Current practice of applying time-resolved PIV is limited to very low jet velocity [8]. Moreover, a reasonable estimation of cavitation can be achieved only as jet flow patterns are fully acknowledged [9].

In this section, the submerged waterjet is produced at high jet pressures of 13, 16 and 18 MPa. Several measurement techniques are used in combination. Local flows near the nozzle are measured with particle image velocimetry. Vorticity distribution is obtained based on the acquired velocity data. Transient variation of cavitation cloud immediately downstream of the nozzle is recorded with high-speed photography. Pressure fluctuations excited by the submerged waterjet stream in the water tank are monitored with miniature pressure transducers.

3.1.1 Experimental Set-Up

3.1.1.1 Experimental Rig

A schematic view of the experimental rig, as well as the optical components, is displayed in Fig. 3.2. The nozzle diameter D is 2.0 mm and the tank is 1800 mm long. Thereby, the submerged waterjet can fully develop in the streamwise direction. A weir plate is deployed near the downstream end of the tank. With this weir plate, the liquid level can be maintained at a certain height, thus the waterjet is exempted from the influence of the submerged depth and free surface oscillation [10]. Here, the streamwise direction is defined as the x direction of a rectangular coordinate system, and the center of the nozzle outlet section is set as the origin of the coordinate system, as indicated in Fig. 3.3.

Tracing particles are required for the implementation of optical flow measurement. With such a small nozzle, provided that the tracing particles are circulated with the bulk flow, they might accumulate in the narrow nozzle passage or emit from the nozzle at a high concentration, thus arousing unpredictable uncertainties. After an evaluation and a preliminary experiment, the fluorescence particles of Rhodamine 6B are seeded in the tank and no tracing particles were deployed in the entire circuit. A homogeneous distribution of the Rhodamine 6B particles is achieved through pre-mixing in the tank. With the optical configuration shown in Fig. 3.2b, a light sheet illuminates the cross section covering the waterjet stream axis and those Rhodamine 6B particles. Images are recorded with a CCD camera and the frame rate is gauged to adapt to the covered area and flow velocity magnitude.

3.1.1.2 Cavitation Observation

Cavitation phenomenon is observed with an Olympus I-SPEED 3 high-speed camera. Meanwhile, an Olympus ILP-2 light source is used to emit light and the light is transmitted through an optical fiber with a diameter of 6 mm. Prior to penetrating into the tank, the light beam passes through a 5 mm thick acrylic diffuser plate which is used to produce a uniform light distribution. In view of the cavitation inception area and the streamwise distance from the nozzle for which cavitation can sustain, a macro lens is used to assist the operation of the high-speed camera. The sampling frequency of the camera is set as 4000 frames per second.

3.1.1.3 Pressure Fluctuation Measurement Apparatus

Pressure fluctuations near the submerged waterjet stream are captured with miniature pressure transducers. The installation of the sensing plate directly facing the jet impact would hamper the progression of the flow, particularly as the jet

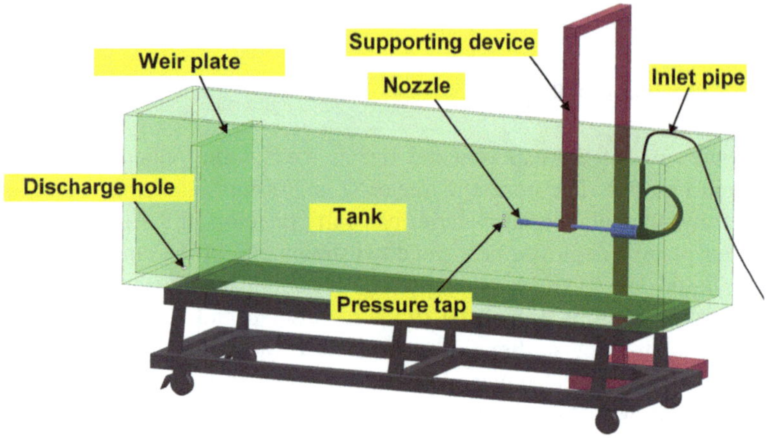

(a) Schematic of the experimental rig

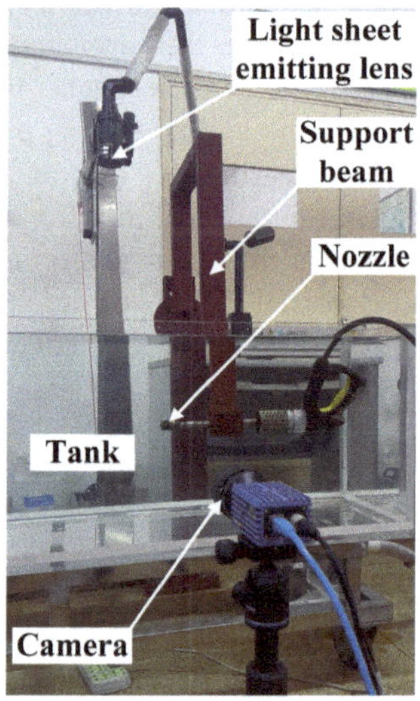

(b) Optical configuration

Fig. 3.2 Submerged waterjet experiment rig

pressure is high [11]. The arrangement of pressure transducers is shown in Fig. 3.4. Among the three identical pressure transducers displayed, the data obtained with the middle one are processed as representative. The relative position between the nozzle and the pressure transducers can be changed through regulating the position of the nozzle. The transverse distance between the jet axis and the middle pressure transducer is set to 200 mm and in streamwise direction, this pressure transducer is set to 50 mm downstream of the nozzle outlet section, a position corresponding to $x/D = 25$. An LMS vibration monitoring system is used to acquire pressure data and

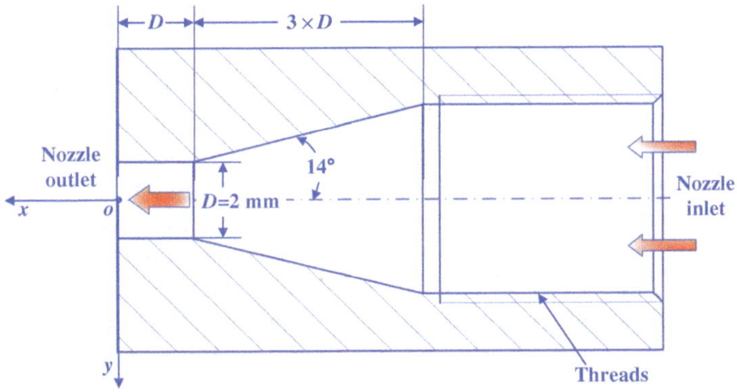

Fig. 3.3 Coordinate system relative to the nozzle

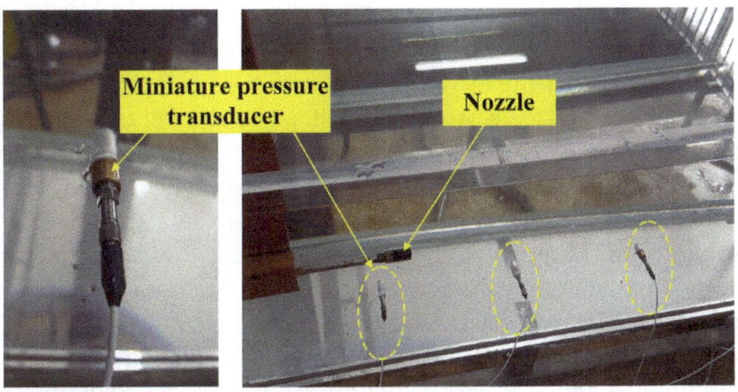

Fig. 3.4 Configuration of pressure transducers

to calculate the typical frequencies and corresponding pressure fluctuation amplitudes. The rise time of the transducers is shorter than 2 μs. The accuracy of the transducers is 0.25%FS.

3.1.2 Average Flow Characteristics of Submerged Waterjet

For the jet pressures considered, the initial segment of the waterjet stream is coherent and the diameter of this segment is equal to the nozzle diameter. This segment is featured by identical velocity magnitudes over its cross-section. The problem with such a segment lies in the reflection of incident light, which leads to statistically unstable results within the range of $x/D < 10$. Even for the non-submerged waterjet, there exists the similar problem [12]. Furthermore, it is possible that consecutive vortex rings dominate the jet stream segment immediately downstream of the nozzle. Therefore, in streamwise direction, data are absent for $x/D < 10$.

The variation of the jet velocity in the streamwise direction reflects the fundamental properties of the jet stream. In Fig. 3.5, velocity distributions over the jet axis in streamwise direction is plotted where U and U_m represent the jet velocity and the maximum jet velocity on each cross section, respectively. Expressed in a non-dimensional manner, the gap between the three curves is narrow. Similarity is evident for the three jet pressures. Reference [13] proposed a theoretical curve indicating the variation of jet velocity, it is also plotted in Fig. 3.5. There is a moderate agreement between the present experimental results and the theoretical data. This is a proof that the waterjet has achieved self-preservation at $x/D > 10$. As for the fluctuation of the jet velocity, it depends on the intense exterior disturbance on the fluid at the jet axis [14].

The variation of jet velocity in transverse direction is illustrated in Fig. 3.6, where $y_{1/2}$ and y stand for semi-width and radial coordinate, respectively. Here, the cross section of $x/D = 15$ serves as a representative of all cross sections beyond $x/D = 10$. It is deemed that the cross sections with a streamwise distance from the nozzle larger than $x/D = 15$ are associated with a higher level of regularity than the cross section of $x/D = 15$. Although jet pressures deviate from each other, highly uniform velocity distributions are seen in Fig. 3.6. Moreover, the three curves are smooth, implying that radial velocity fluctuations are less intense than those in streamwise direction. In terms of the variation tendency, experimentally obtained distribution in Fig. 3.6 is in good agreement with the result of direct numerical simulation (DNS) [15].

Semi-width is a parameter that represents the demarcation between high- and low-velocity fluids in the waterjet stream. From another viewpoint, the semi-width curve is a boundary for concentrated energy and an indication of where high

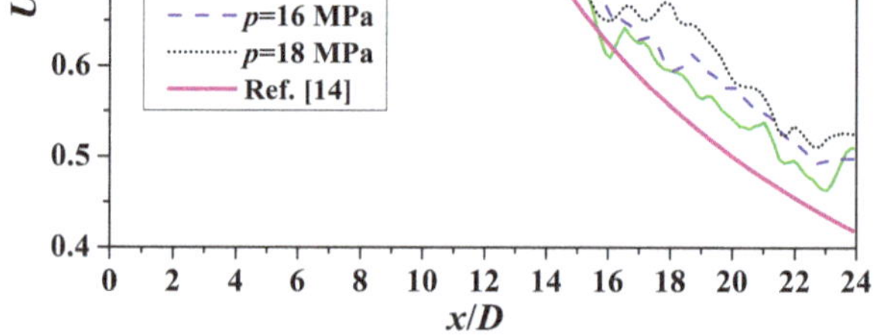

Fig. 3.5 Velocity distributions in the submerged waterjet at different jet pressures

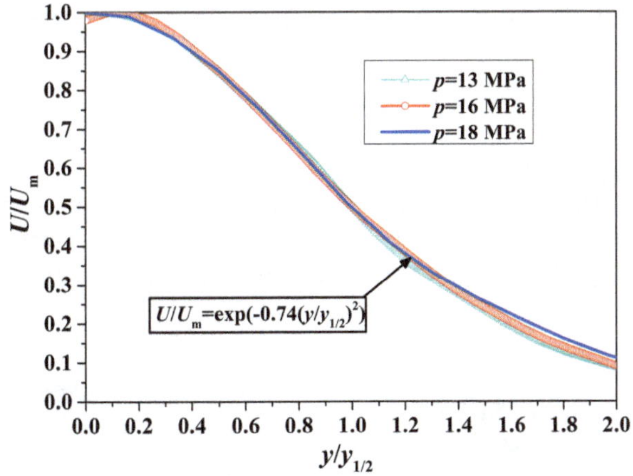

Fig. 3.6 Variation of waterjet velocity in transverse direction at x/D = 15

velocity gradients arise in the submerged waterjet. Ambient water might have invaded into the waterjet stream outside of the semi-width curve. In Fig. 3.7, the variation in the non-dimensional semi-width in streamwise direction is plotted for the three jet pressures considered. It is seen that all the three curves do not conform to linear relationship, but the three curves are similar in terms of variation tendency and the magnitudes. In addition, the three curves can be approximated by a straight line with a slope of 0.065 with the coordinates adopted.

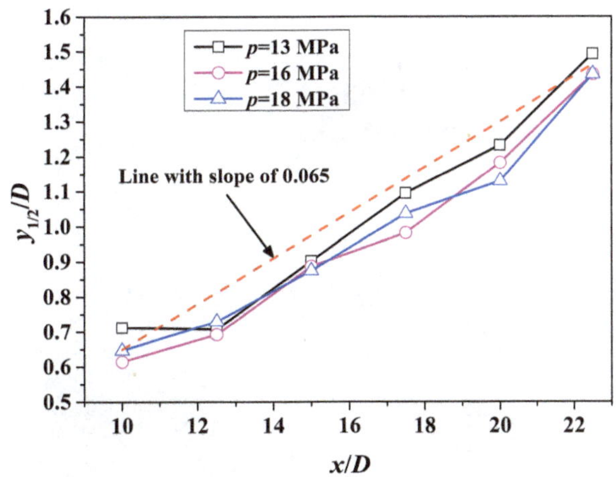

Fig. 3.7 Comparison between semi-width curves at different jet pressures

3.1.3 Vorticity Distribution and Pressure Fluctuation

3.1.3.1 Vorticity Distribution

Temporal vorticity distributions at the three jet pressures are shown in Fig. 3.8. Vorticity vectors are calculated based on the transient velocity distribution. Only vorticity magnitude is given in Fig. 3.8 and the waterjet advances from left to right. Regarding the instantaneous distribution, vorticity magnitudes are approximately

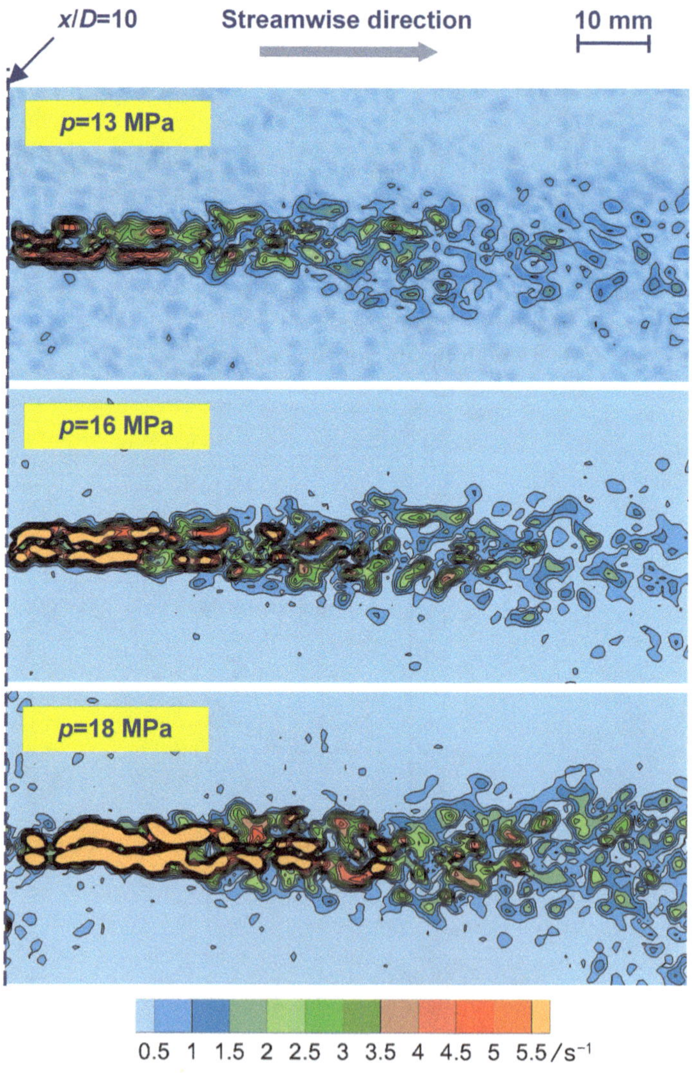

Fig. 3.8 Temporal vorticity distributions at different jet pressures

symmetric with respect to the jet axis, as is shared by the three situations. Meanwhile, as jet progresses, vorticity magnitude decreases and eventually, small vorticity elements are prevalent in the waterjet stream. Vorticity originates from velocity gradients and is also an essential factor underlying flow structures. Downstream vorticity distributions displayed in Fig. 3.8 are related to the collapse of the integrated waterjet stream.

As the jet pressure increases, vorticity magnitudes near the nozzle outlet increase apparently. Since the nozzle outlet diameter is only 2.0 mm, the upper and lower high-vorticity bands immediately downstream of the nozzle interact with each other. The production of small flow structures near the jet axis at nozzle outlet is thereby facilitated.

In view of the temporal vorticity distributions in Fig. 3.8, they are similar except for the magnitude. To describe the vorticity distribution in an averaged manner, at $p = 18$ MPa, cross-sectional averaged vorticity distributions are displayed in Fig. 3.9, where ω and ω_m are vorticity and the maximum vorticity magnitude at each cross section, respectively. Here, vorticity signs are expressed as well. The three curves were constructed based on 1024 transient images. There are two peaks symmetrically distributed over the jet axis, $y/D = 0$. The value of y/D corresponding to the peak deviates from the centerline with the jet development. With the jet pressure and nozzle diameter considered here, the radial distance between the two peaks is rather small [16]. The amplitudes of the two peaks are nearly equivalent. As x/D increases, the concentration of high vorticity is impaired progressively. Meanwhile, away from the jet axis in radial direction, vorticity vanishes finally. Even with the nozzle of square outlet cross section, the dual peaks are obvious [17].

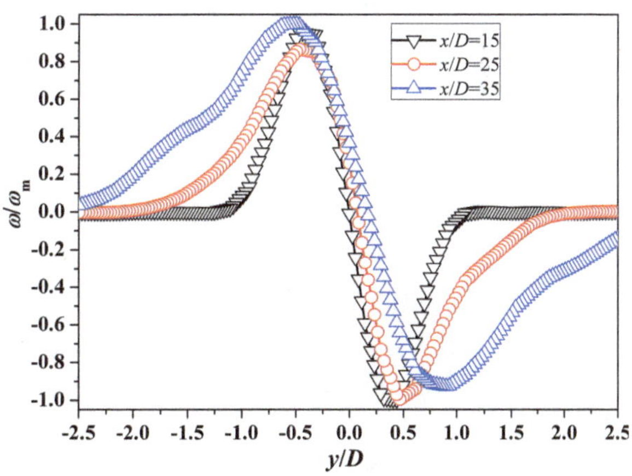

Fig. 3.9 Vorticity distribution in the submerged waterjet at p = 18 MPa

3.1.3.2 Pressure Fluctuation

The raw pressure data obtained with the miniature pressure transducers are time dependent. Here, the fast Fourier transformation (FFT) method is used to extract characteristic frequencies and pressure fluctuation amplitudes in frequency domain. Typical frequency spectra associated with the three jet pressures are shown in Fig. 3.10 where the frequencies range from 0 to 315 Hz. It is seen that low-frequency components are salient. The emergence of low excited frequencies is ascribed to the evolution of large-scale flow structures at jet edge, as is substantiated through the proper orthogonal decomposition (POD) results [18]. The influence of the plunger pump frequency on the low-frequency components in the submerged waterjet has been confirmed in [19]. Here, since that the monitored point is in the water tank, the properties of the low-frequency pressure signals are determined by both output pressure of the plunger pump and the flow structures near the nozzle. From $p = 13$ to 18 MPa, the pump frequency changes from 18 to 25 Hz. In Fig. 3.10, the pump frequencies, accompanied with moderately high pressure fluctuation amplitudes, are recognized. At each jet pressure, there is a most predominant frequency which is defined as the frequency at which the maximum pressure fluctuation amplitude arises. The most predominant frequency increases with the jet pressure but is still limited in the low frequency range. Such a frequency is related to the flow structures excited by the jet stream in the water tank.

The overall variation tendency of the pressure fluctuation amplitude is that the larger the jet pressure, the higher the pressure fluctuation amplitude. Although a fairly large water tank is used here to accommodate the submerged waterjet, the boundary effect is non-neglectable. In general, the tank wall facilitates the

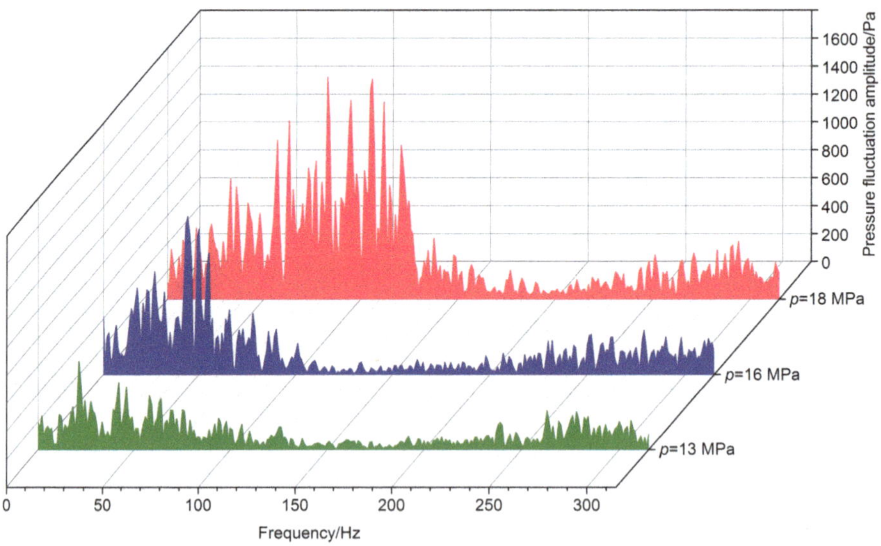

Fig. 3.10 Pressure fluctuations in frequency domain

generation of large-scale flow structures [20]. Moreover, at p = 18 MPa, around the frequency of 100 Hz, high pressure fluctuation amplitudes are abundant. This implies that flow structures with diverse scales exist in the water tank. Pressure fluctuations associated with the vortex ring cavitation have been identified in [21]. In Fig. 3.10, the emergence of those high-frequency peaks is ascribed to cavitation. Moreover, the effect of cavity geometry and the traveling velocity of the cavity on high-frequency parts is perceptible.

At each jet pressure, a statistical analysis is performed based upon a data group containing 512 frequency spectra. Consequently, the maximum, minimum and mean pressure fluctuation amplitudes are obtained. Both the maximum and mean pressure fluctuation amplitudes are plotted in Fig. 3.11. As the jet pressure increases, there is a constant increase of the maximum pressure fluctuation amplitude. Nevertheless, the mean pressure fluctuation amplitude only exhibits a relatively moderate rise. It means that the entire spectrum involves a wide range of low pressure fluctuations.

The three maximum pressure amplitudes correspond to three dominant frequencies, f_d, respectively. Meanwhile, the jet pressure is expressed by the subscript. As the jet pressure increases, the dominant frequency increases, which demonstrates that fast-varying flow structures are influential at high jet pressures.

3.1.4 Cavitation

Cavitation phenomenon in the submerged waterjet bears unique unsteady characteristics. As for the cavitation inception mechanism, available knowledge is far from sufficient. With the high-speed camera, a macro lens and LED light source, cavity profile immediately downstream of the nozzle can be captured. In this context, cavitation profiles corresponding to the three jet pressures are shown in Fig. 3.12 where jet develops from right to left. Here, T_{0A}, T_{0B} and T_{0C} represent individual starting moments relative to the subsequent images.

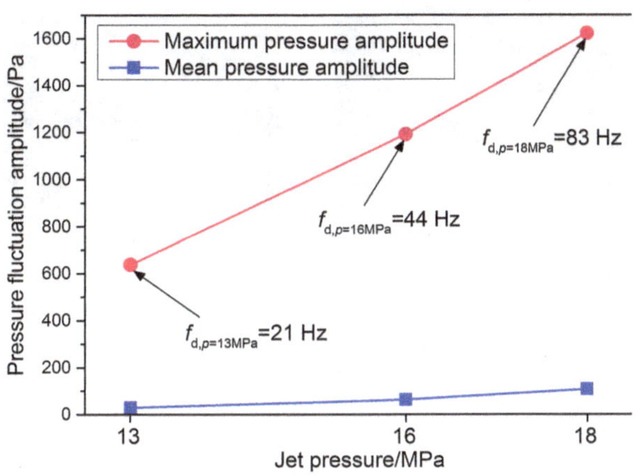

Fig. 3.11 Statistical pressure fluctuations

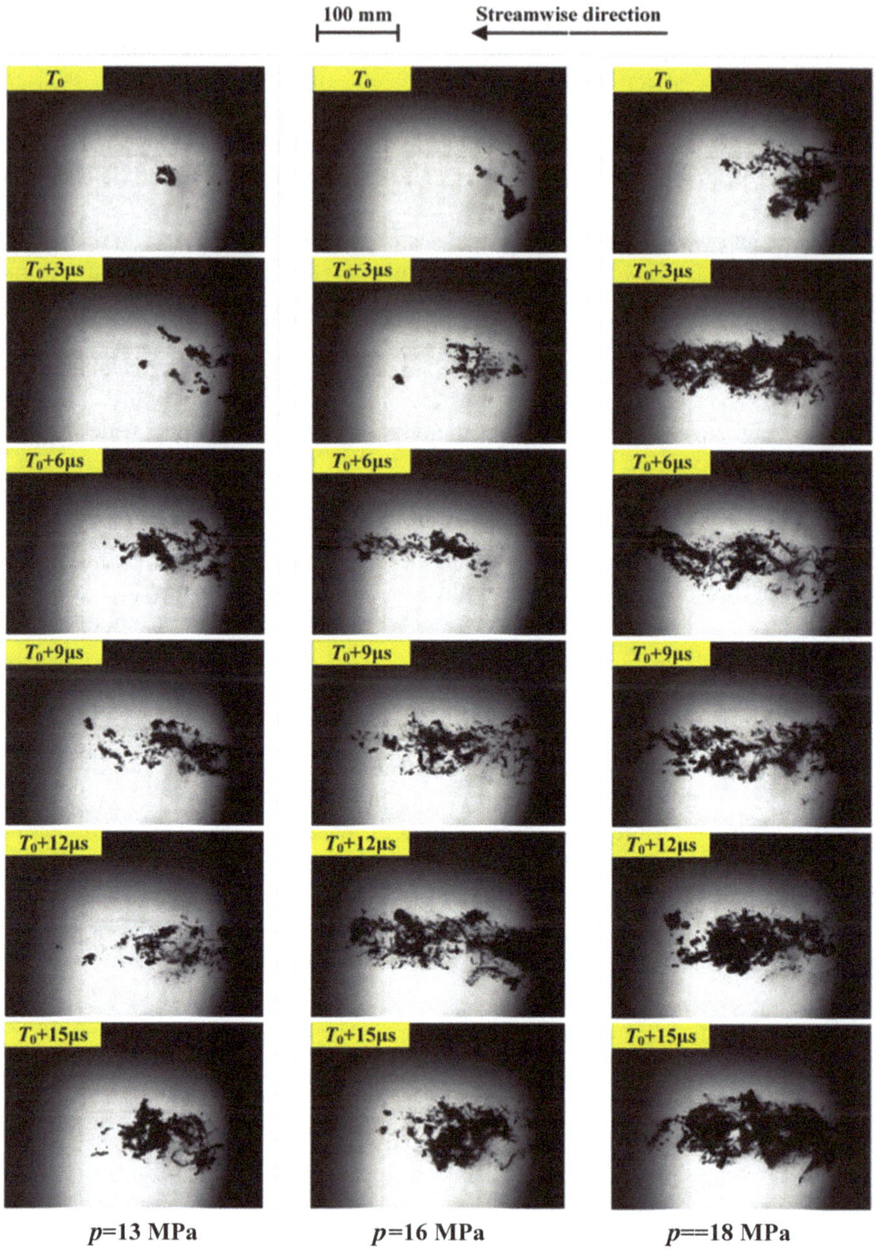

Fig. 3.12 Evolution of cavitation clouds at three jet pressures

The evolution of cavitation evolution is an intermittent process during which cavity profiles vary considerably. At a certain jet pressure, the development of cavitation is composed of considerably different consecutive cavity images. The evolution of cavitation is bound to trigger multiple frequencies [22]. Furthermore, cavities are entrapped in the main stream and progress downstream along a meandering route. Growing and shrinking of cavitation cloud occur alternately.

At p = 13 MPa, cavitation is characterized by small and dispersedly distributed cavitation elements, and these elements behave as wavy structures. The overall cavitation area fraction is small with respect to the whole image. Regarding the progression of the cavitation elements, they match the temporal vorticity distributions instead of averaged flow parameter distributions. Therefore, it is inferable that cavitation inception is inherently related to local vorticity. As the jet pressure increases, cavitation is reinforced and cavitation cloud takes shape. The increase in the jet pressure results in the decrease of cavitation number, which can also be realized by lowering the static pressure of ambient water [23]. Meantime, cavity cloud can sustain for a fairly long streamwise distance. In particular, at p = 18 MPa, the cavitation cloud is coherent and assumes a considerable area fraction in the captured images. The correspondence is also implied in Fig. 3.8, where the streamwise stretching of high-vorticity bands is apparent as the jet pressure increases.

Small temporal scales are associated with the typical cavity patterns shown in Fig. 3.12, as motivates a statistical investigation of the cavitation evolution based on consecutively captured images. An attempt has been made to calculate the shedding frequency through characteristic scale [24]. Here, it is obvious that the variation of the transverse cavitation cloud width is difficult to measure. Thus, the variation of the streamwise cavitation cloud length during image acquisition is employed to illustrate the cavitation evolution. An image set containing 600 consecutively captured images is selected for each case. A preliminary statistical investigation proves that the variation frequency of cavitation cloud is unambiguous and traceable. Typical variation of cavitation cloud length at each jet pressure is illustrated in Fig. 3.13.

Cavitation evolution is a multiple-frequency process, as indicated in Fig. 3.13. In view of the time-varying cavitation cloud length, three sequentially appeared phenomena, namely cavitation cloud extending, cavitation cloud preserving and cavitation cloud shrinking are recognized, as denoted in Fig. 3.13. At different jet pressures, the time span and extent associated with the same attribute are different. At p = 13 MPa, the downstream extension of cavitation cloud is weak. In contrast, at p = 18 MPa, long cavitation clouds are prevalent, as is in good accordance with Fig. 3.12. Furthermore, the cavitation cloud length varies at p = 18 MPa and the time intervals between two neighboring highest cloud length peaks is short relative to the other two cases. At low jet pressure, a long period is consumed by the transition from a minimum cavitation cloud length to another, in particular at p = 13 MPa. In this connection, the jet pressure serves as the primary impetus behind cavitation strengthening, irrespective of the time span [25]. Additionally, at higher jet pressure, cavitation patterns are more consistent in terms of spatial and temporal features.

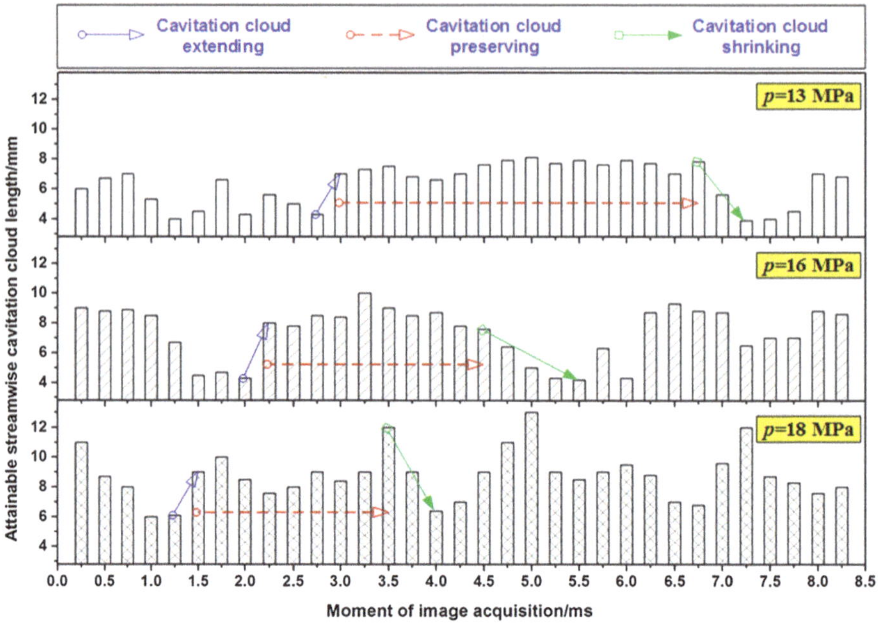

Fig. 3.13 Typical time-dependent variation of attainable streamwise cavitation cloud length

The time-dependent variation of the cavitation cloud length is depicted in Fig. 3.13. A combination of the three consecutive phenomena of cavitation cloud extending, cavitation cloud preserving and cavitation cloud shrinking serves as an integral unit, then the variation frequency of the unit is calculated. From $p = 13$ to 18 MPa, the variation frequency increases from 222 to 364 Hz. With a further examination over the whole sampled data set, these calculated frequencies are statistically repeatable. In this connection, with reference to the frequency spectra shown in Fig. 3.10, the excited high frequencies over 200 Hz are noticeable. High-frequency components are reinforced as the jet pressure increases. Such a variation tendency is in accordance with that indicated in Fig. 3.13.

With this case, it is perceivable that the submerged waterjet faces considerable resistance from ambient water at high jet pressures. Jet velocity attenuates rapidly in streamwise direction. It is interesting that dual-peak vorticity distribution arises near the nozzle outlet. As the jet pressure increases, coherent cavitation clouds are formed and the area fraction assumed by cavities increases as well. The instantaneous cavitation cloud profiles are compatible with temporal vorticity distributions. Cavitation evolution encompasses cloud extending, cavitation cloud preserving and cavitation cloud shrinking. Pressure fluctuations near the submerged waterjet stream bear explicit characteristics of flow excitation. The frequency spectra are dominated by low-frequency components. The effect of cavitation is testified by relatively high frequencies in the pressure fluctuation spectra.

3.2 Submerged Waterjet Issued at Ultra-High Jet Pressures

To illuminate primary factors influencing the morphology of the surface impinged by the submerged waterjet, experiments are performed at jet pressures from 200 to 320 MPa. In general, the jet pressure exceeding 276 MPa is defined as ultra-high jet pressure. For these extreme jet pressures, both the flow characteristics and the cavitation phenomenon deserve much attention. In consideration of the shortages of existing measurement techniques, we use the target specimen as the sensor of cavitation. Here, copper alloy specimens are used to endure the impingement of the waterjet. The microhardness of the specimen is measured. Surface morphology is observed using an optical profiling microscope. Meanwhile, pressure fluctuations near the jet stream are acquired with miniature pressure transducers, which is similar to the technique used in Sect. 3.1.

Provided that a solid specimen is impacted by the submerged waterjet, the explosion of cavitation bubbles produces an enormous force onto the specimen surface, as is an important factor that leads to surface strengthening, surface erosion or even surface damage [26]. Previous studies are dedicated to waterjets at medium or low jet pressures. In comparison, submerged waterjets at pressures over 200 MPa have rarely been investigated.

Numerical simulation of cavitation bubble generation and collapse is not an easy task [27]. In terms of cavitation erosion simulation, it is even complicated since that stress propagation in the specimen and surface damage exposed to repeated cavitation impact are lack of mechanism illustration [28]. At present, cavitation erosion can only be approximately predicted based on the cavitation inception position and cavity profiles [29]. Although the joint application of CFD and finite element method (FEM) has been attempted to simultaneously treat the jet flow and the solid specimen impacted, the results need validation [30].

Experimental techniques for measuring cavitation erosions advance rapidly in recent years, and optical methods are prevalent now [31]. Here, it is noteworthy that flow visualization is difficult to accomplish under high jet pressure conditions. Regarding flow velocity measurement, the resolution of the commonly used optical instruments cannot fulfill the requirements.

3.2.1 Cavitation Erosion Mechanism

As cavitation bubble collapse acts repeatedly upon the target surface, fracture failure might occur. There are two influential factors in this process, one is the bubble distribution over the cross section of the waterjet stream, and the other one is bubble velocity. Essentially, both factors are difficult to measure with current techniques. As bubbles progress with high velocity, they arrive at the target surface with considerably high kinetic energy, which constitutes another impact mechanism that cannot be neglected. Hitherto, the identification of these factors has not been accomplished yet. This issue is also related to material properties, as enhances the difficulty level of relevant studies.

3.2.2 Experimental Methods and Rig

The experimental rig is established on the platform of an ultra-high jet pressure generation system and a base was specifically designed to support and fix the test specimen. In view of the restriction of the power consumed and the flow rate of the discharged water, three nozzle diameters, namely 0.006, 0.007 and 0.009 inches are selected. With the size of the submerged waterjet stream and the jet pressures considered, it is impractical to manufacture a plexiglass tank or pipe to conduct jet and cavitation visualization experiments [32]. Meanwhile, since the nozzle outlet diameter is small, the integrity of the waterjet stream can be remained for a fairly long streamwise distance as it leaves the jet nozzle [33].

As shown in Fig. 3.14a, the submerged water jet direction is vertical to the target surface. Both the waterjet pressure and the standoff distance between the nozzle exit and the specimen surface are considered. An AxioCSM 700 confocal microscope is used to observe cavitation erosion pits and surface roughness. The microscope is a true color instrument and the resolution attains 0.16 μm. It operates with reflected lights to measure precisely the topographical parameters of the impacted surface. The local height that can be measured with the microscope ranges from 20 nm to several millimeters. The MATLAB code is used to batch process the image data. The surface morphological features will be described through local height or surface roughness of the impinged surfaces. With image analysis, quantitative information can be extracted from images obtained with the scanning electron microscope (SEM) [34]. Relative to SEM, the measure used here is straightforward and facilitates the comparison of local surface morphology.

Apart from surface morphology, pressure fluctuations are measured using eight miniature pressure transducers, as shown in Fig. 3.14b. It should be pointed out that the pressure measured here is transient pressure near the submerged waterjet stream instead of the pressure sensed by the specimen [11]. Such a scheme is utilized to detect pressure fluctuations associated with cavitation and thereby cavitation erosion is traced from flow aspect.

The test rig standard of jet cavitation erosion is issued by ASTM and the major purpose of the standard is to test the anti-cavitation capability of materials [35]. In this study, the cavitation erosion is used to assess cavitation, so operation parameters that influence the waterjet stream profile and velocity are emphasized.

Material property determines to a large extent the response of the specimen to waterjet impact [36]. Copper alloy specimens with identical dimensions, the thickness of 4.8 mm and the diameter of 40.0 mm, are selected as the impinged targets, as shown in Fig. 3.15. Prior to the test, the specimens are processed with mechanical polishing technique to remove surface contaminations and defects.

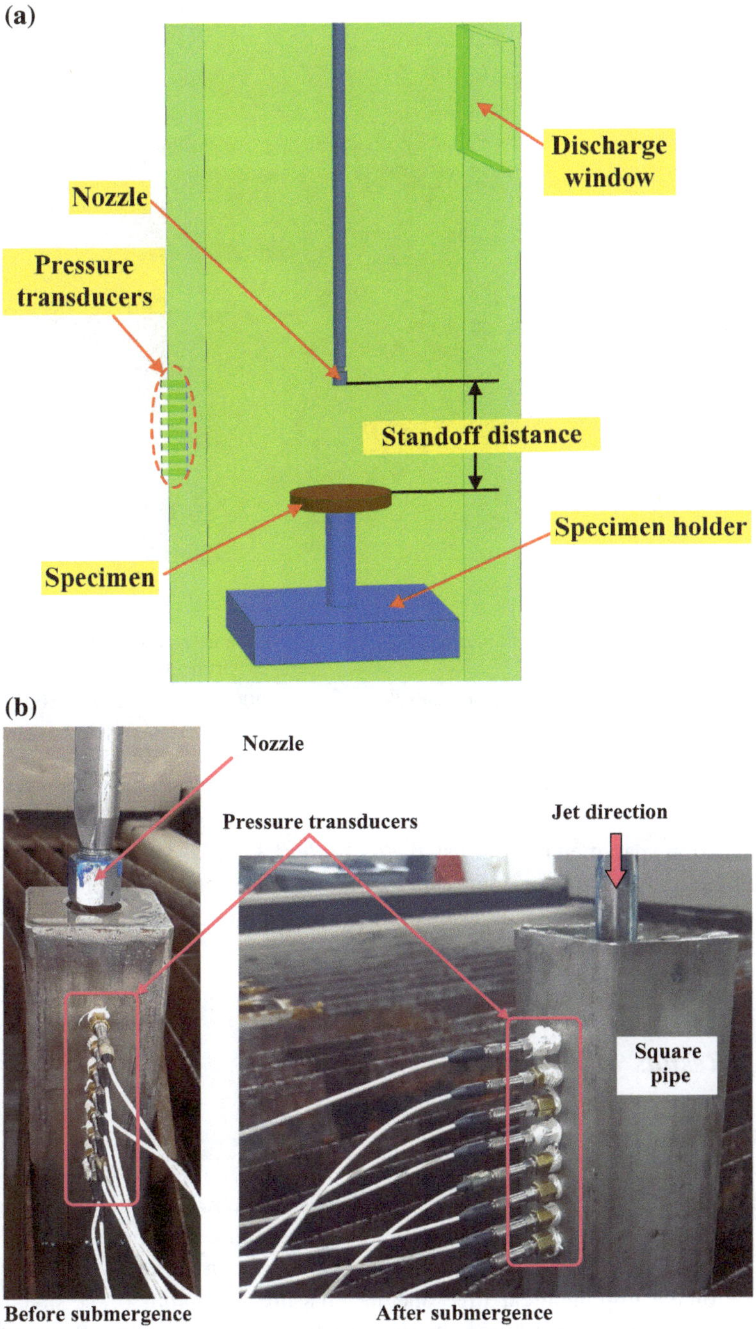

Fig. 3.14 Experimental rig and nozzles: **a** Schematic view of the experimental rig, **b** configuration of miniature pressure transducers

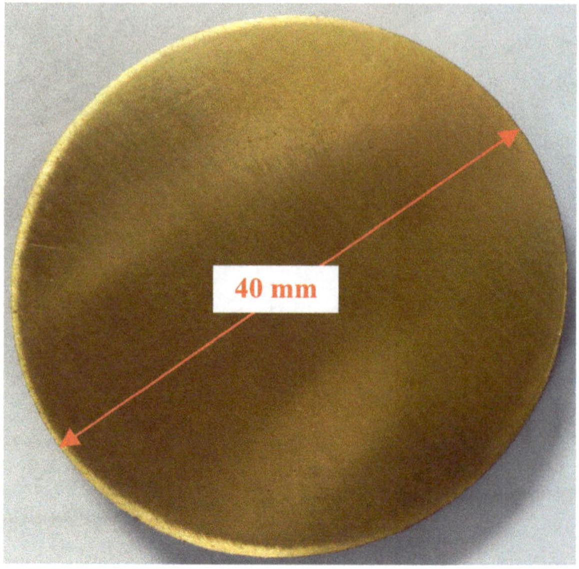

Fig. 3.15 Test specimen

Table 3.1 Operation parameters

Parameter	Parameter values used in the experiment				
Jet pressure /MPa	200	260	280	300	320
Standoff distance /mm	5	9	15	25	35
Impingement time /min	2	3	5	7	9

The operation parameters are listed in Table 3.1. To implement comparison between various operation conditions, various combinations of the three parameters are used, which means that one jet pressure is associated with different standoff distances and impingement time.

3.2.3 Results and Analysis

3.2.3.1 Microhardness

Microhardness is measured with a Vickers hardness tester. The initial microhardness of the specimens is measured prior to the impingement experiment, and the values range from 89 to 91 HV. During the experiment, with different specimens, the influence of the impingement time and jet pressure is examined. To quantify the influence of the impingement time, the standoff distance between the nozzle outlet and the target surface is fixed at 15 mm. The variation of microhardness with impingement time is plotted in Fig. 3.16 where the jet pressure of 280 MPa is kept constant.

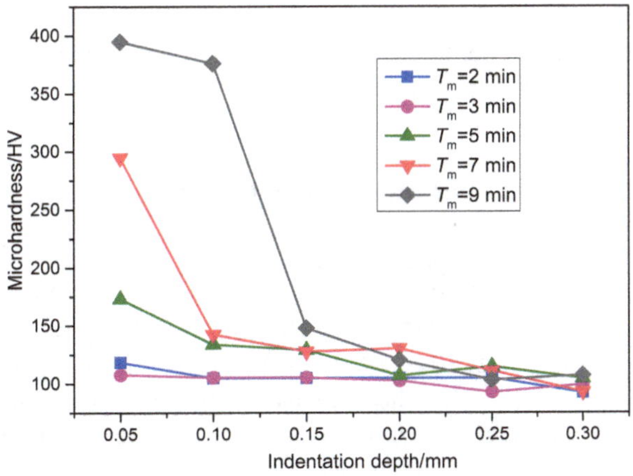

Fig. 3.16 Variation of microhardness with impingement time at 280 MPa

At small indentation depth, the microhardness is distinctly high, which is seen as the impingement time exceeds 2 min. For the two cases with impingement time of 2 and 3 min, the microhardness curves are rather flat, and the gap between the two curves is not clear, indicating an incubation stage of cavitation damage. As the impingement is advanced, an apparent tendency is that long impingement time results in high hardness. Meanwhile, at initial stage, microhardness drops sharply as the indentation depth increases. It implies that the impact effect of the submerged waterjet concentrates in a shallow layer underneath the target surface. The thickness of the layer increases with the impingement time. The result in Fig. 3.16 conforms to the reported variation tendency of microhardness [37].

As the jet pressure varies, the variation of microhardness is measured and the result is plotted in Fig. 3.17, and the impingement time of 3 min is constant. As the jet pressure rises, a consistent increase in the maximum microhardness is seen.

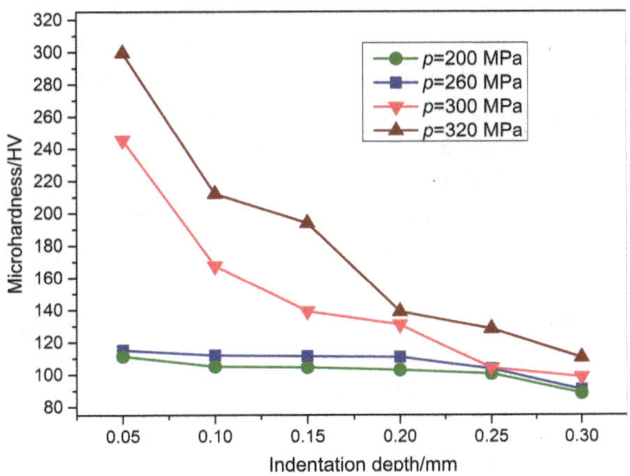

Fig. 3.17 Variation of microhardness with jet pressure after impingement of 3 min

Similar to Fig. 3.16, the maximum microhardness emerges at the minimum indentation depth. As the jet pressure increases further, high microhardness penetrates deep into the specimen.

3.2.3.2 Surface Morphology Feature

To identify morphological features of the impinged specimen surface, an amplification factor of 50 is set for the microscope; thereby the accurate dimensions of the sample area are 234 μm × 188 μm. According to such a setting, the resolution of the image is 1280 × 1024 pixels. The surface morphological pictures are displayed in Fig. 3.18. With the dimensions of the sampled surface area, it is impossible to formulate a global view, so only typical surface features are illustrated. In Fig. 3.18, each column of subfigures corresponds to a certain jet pressure, as marked at the bottom of each column. Regarding each column, the five images are sequentially arranged from the top side to the bottom side. The five subfigures in each column are captured at impingement time of 2, 3, 5, 7 and 9 min, respectively.

It should be noted that the erosion pits do not appear at the position where the jet axis intersects with the target surface, this conforms to the general knowledge that it is the vortex ring surrounding the jet axis that nurtures cavitation bubbles [38]. It has been concluded that the impact pressure will successively experience two peaks as the waterjet progresses, the first one is due to the jet impact and the subsequent one results from the collapse of the bubble ring [4]. In view of the exceedingly high jet pressures adopted here, any of the two peaks imposes a potential threat to the integrity of the target surface.

Waterjet impingement process is dominated by multiple factors. High jet pressures lead to high impact energy, which is associated with the stagnation pressure as the waterjet reaches the target surface. The plastic deformation in the impinged surface depends on the jet pressure. Meanwhile, jet energy will dissipate rapidly with the increase in the standoff distance. At high jet pressures, cavitation is expected be much intense [39]. Consequently, the surface damage due to cavitation erosion will be severe relative to that at low jet pressures [40]. The process of cavitation erosion can be divided into four successive stages. The first stage is featured by few small pits, which are accompanied by surface deformation, as is similar to the general description of the initial stage of cavitation erosion [41]. At the second stage, both the pit size and pit number increase. But those pits are still sparsely distributed. At the third stage, few pits with large size are found but the majority of the pits are featured by small size. Meanwhile, large-size pits are associated with irregular peripheries. The dimensions of large erosion pits are affected by the magnitude of the bubble explosion load, the load duration period and the scope covered by the load [42]. Regarding the fourth stage, large-size pits are apparent and some of them are apparently connected. It can be predicted that as cavitation erosion develops further, large material elements will be removed from the specimen surface but there exists a threshold amount material removal which signifies a balance between the impact of the submerged waterjet and the response of the specimen.

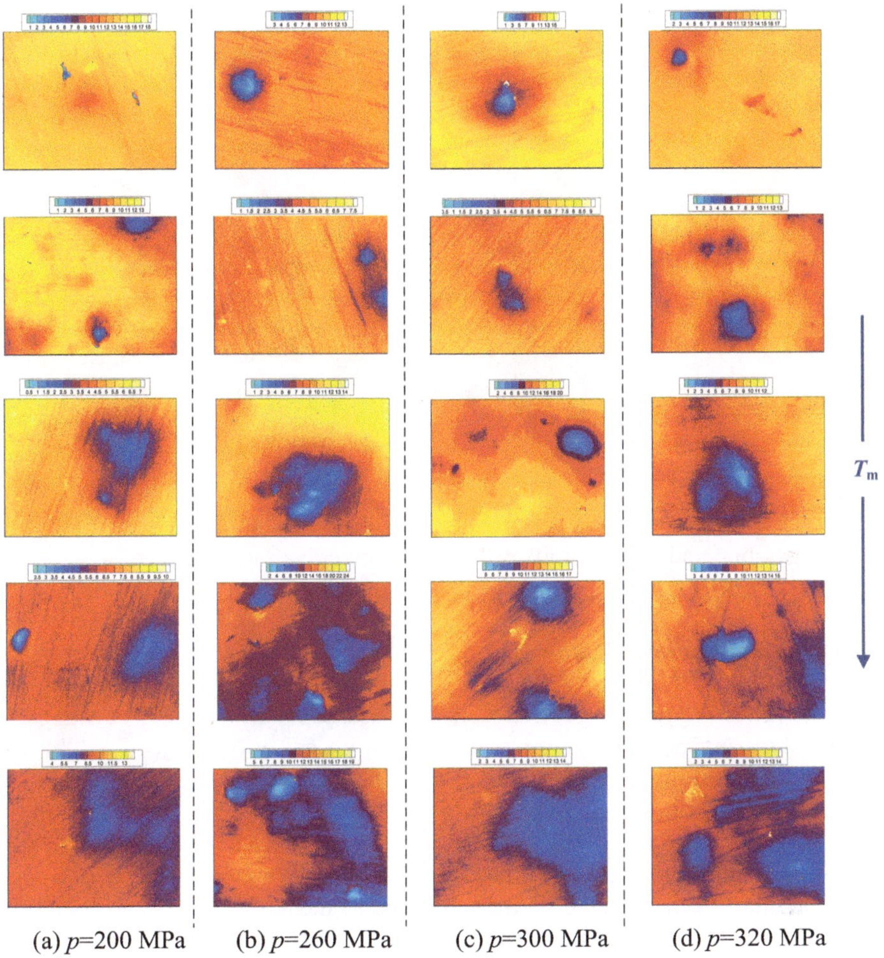

Fig. 3.18 Surface morphological features for various jet pressures

3.2.3.3 Cavity Dimension Analysis

In Fig. 3.18, cavity pit shapes are diverse and irregular, which is ascribed to the water wedge effect and the shock wave resulting from cavitation bubble collapse. Here, the diameter of the circle that can fully accommodate the cavity rim is used to express cavity size. Such an approach considers not just the profiles of the erosion pits but also the irregularity of the pit edge. Based upon the captured images, cavity size is statistically calculated. According to the explanation of the four stages in Fig. 3.18, the cavity size calculated is categorized and the result is schematically shown in Fig. 3.19. Hitherto, the definition of the stages encompassed by the cavitation erosion process is overwhelmingly qualitative [43]. Figure 3.19 furnishes a comparison of

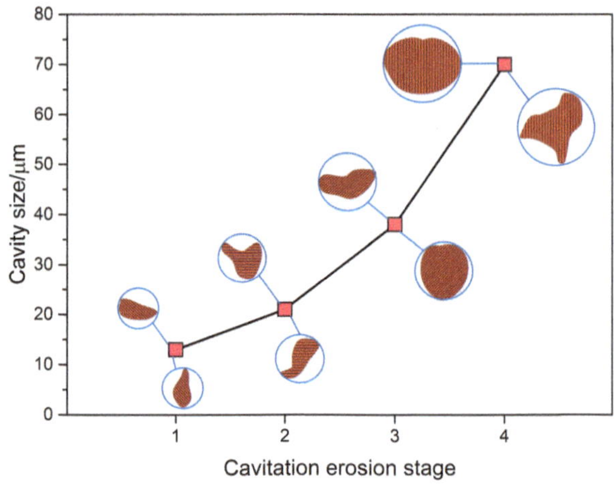

Fig. 3.19 Geometrical property of cavity at different cavitation erosion stages

cavity size between different stages. It is explicit that the cavity size increases gradually at first and then sharply as the cavitation erosion is enhanced.

3.2.3.4 Mass Removal Rate

Based on the above results, generalizable conclusions cannot be easily constructed. Meanwhile, a sufficiently large group of sampled images or data is required. In this context, more information is anticipated to be extracted from the specimens. Macroscopic quantities such as the mass removal rate, defined as the ratio of mass removed to the original specimen mass, and mean erosion pit diameter are frequently adopted to assess jet erosion capability [44]. Weighing method is a conventional and effective method. After the impingement, the sample is dried and weighed. With this method, the influence of the impingement time and the standoff distance can be examined. The variation of the mass removal rate with the two operation parameters is plotted in Figs. 3.20 and 3.21.

It is seen in Fig. 3.20 that the mass removal rate rises consistently with the jet pressure and the impingement time. Nevertheless, the imparity between the four curves is remarkable. For short impingement time, the mass removal rate is low even at ultra-high jet pressures. This proves that the cavitation erosion is undeveloped and is in the incubation stage. It can be predicted that energy leading to surface damage is accumulated in this phase. As the impingement time increases from 15 to 22 min, a drastic improvement of the mass removal rate is indicated, and the contribution of those large-size erosion pits is phenomenal. In this connection, the situation witnessed here provides a support for the demarcation of cavitation erosion stages with the amount of the mass removed [45]. Since mass removal rate is closely related to the cavitation erosion pit size, the similarity between Figs. 3.19 and 3.20 is appreciable. In addition, a non-dimensional relationship between the

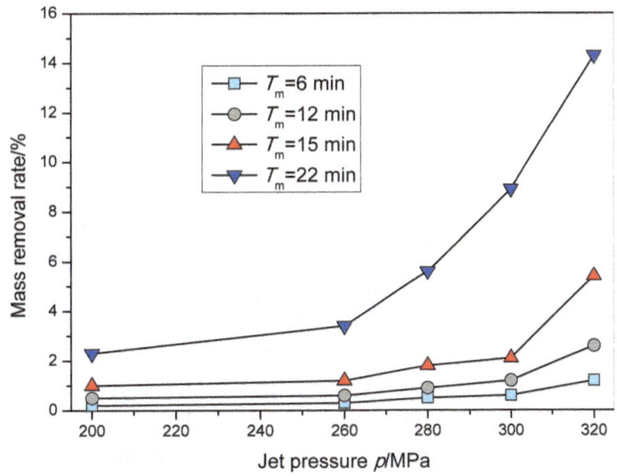

Fig. 3.20 Mass removal rate at different impingement time

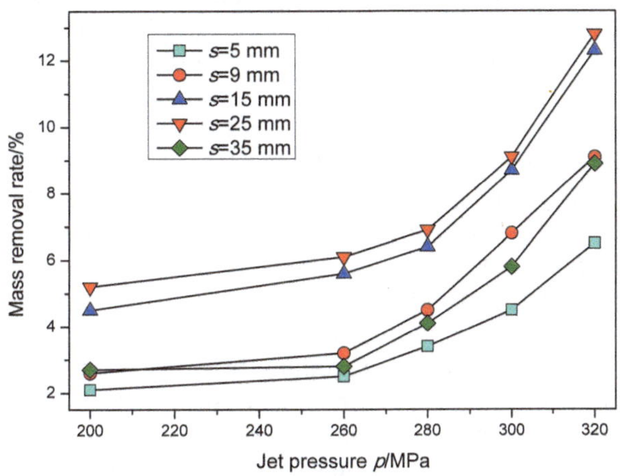

Fig. 3.21 Mass removal rate at different standoff distance

characteristic erosion depth and characteristic impingement time is proposed in [46]. In view of the results shown in Fig. 3.20, such a relationship is only suitable for long-period waterjet impingement.

In Fig. 3.21, the five curves share analogous overall variation tendency of mass removal rate as the jet pressure increases. For certain standoff distance, the mass removal rate increases with the jet pressure. Between jet pressures of 200 and 260 MPa, the upward tendency is mitigated. As the jet pressure increases further, the augment of the mass removal rate is intensified. In Fig. 3.21, at each jet pressure, it is remarkable that the shortest standoff distance is uniformly associated with the lowest mass removal rate. There is an optimum standoff distance for each jet pressure. This situation agrees with the conclusions about the optimum standoff distance at low jet pressures [47]. At very low jet pressures, an approximate slope of the curve representing the variation of the erosion pit size with the jet velocity has been proposed in [48]. Nevertheless, in Fig. 3.21, within the jet pressure range considered, for each curve, the curve slope varies to a large extent with the jet pressure.

3.2.3.5 Pressure Fluctuations

Pressure fluctuations near the submerged waterjet stream bear the effect of cavitation excitation. Furthermore, such an excitation has an essential connection with repeated impact of cavitation bubble explosion on the target surface. With the eight miniature pressure transducers shown in Fig. 3.14, transiently varied pressure is recorded at different vertical positions. Temporal pressure signals are then transformed using FFT approach, and the pressure amplitude spectra are obtained thereby. The vertical position of the nozzle outlet is denoted by Z_0, a constant. Accordingly, the vertical position of the pressure transducer is denoted by Z, which increases with the progression of the waterjet. Pressure fluctuations in the frequency domain at four vertical positions are displayed in Fig. 3.22. In this connection, four jet pressures, 200, 260, 300 and 320 MPa, are selected for the comparison.

It is seen from Fig. 3.22 that low-frequency components are predominant, as is in accordance with typical result of sound pressure level (SPL) obtained with the hydrophone [49]. Meanwhile, as the jet pressure changes, the characteristic frequency corresponding to the highest pressure amplitude peak varies as well. It is generally deemed that as the jet develops downstream, the pressure fluctuation amplitude decreases [50]. Such an overall trend is proved as far as the high jet pressure magnitudes are concerned. As the waterjet progresses from Fig. 3.22a–d, overall pressure fluctuations decay.

In each frequency spectrum, two frequency segments with saliently high pressure amplitudes are seen. One segment is featured by a protruding pressure peak and is related to low frequencies. The other segment exhibits a pressure hump, and the corresponding frequencies are relatively high. It is concluded that the first segment is caused by bubble explosion and the second segment by vortex ring [21]. Such a viewpoint is reasonable at medium jet pressure. For the jet pressure range considered here, the first segment is inseparable from the disturbance of the submerged waterjet to the ambient water, so pressure fluctuation amplitude can maintain for a long streamwise distance, and in Fig. 3.22d, it drops considerably. The second segment depends on the cavitation bubble explosion, and further downstream, the second segment tends to concentrate in a certain frequency range. In Fig. 3.22d, the deviation of the second segment from the first segment is appreciable. Exposed to large magnitudes of jet pressure, the submerged waterjet remains its integrity for a long streamwise distance and cavitation bubbles are abundant. Consequently, the effect of cavitation erosion is enhanced as the jet pressure increases, which is compatible with the surface morphology patterns illustrated in Fig. 3.18.

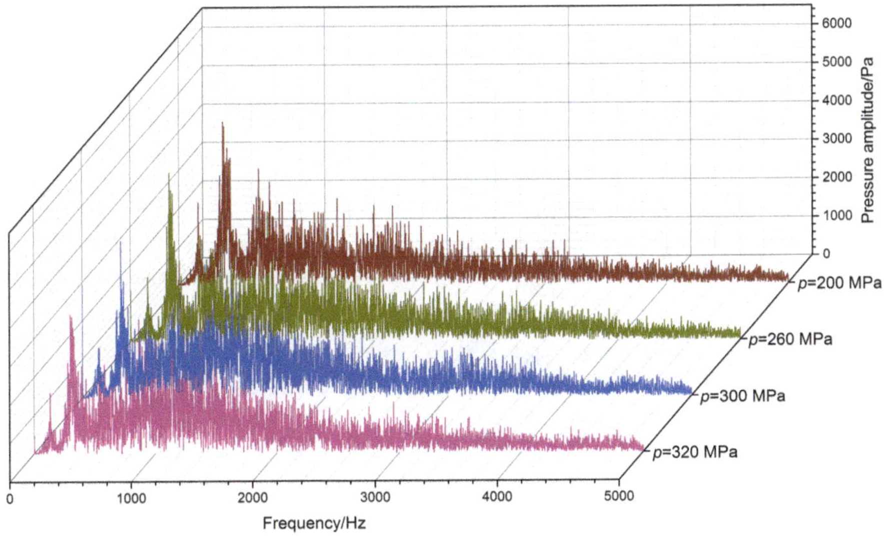

(a) $Z=Z_0$

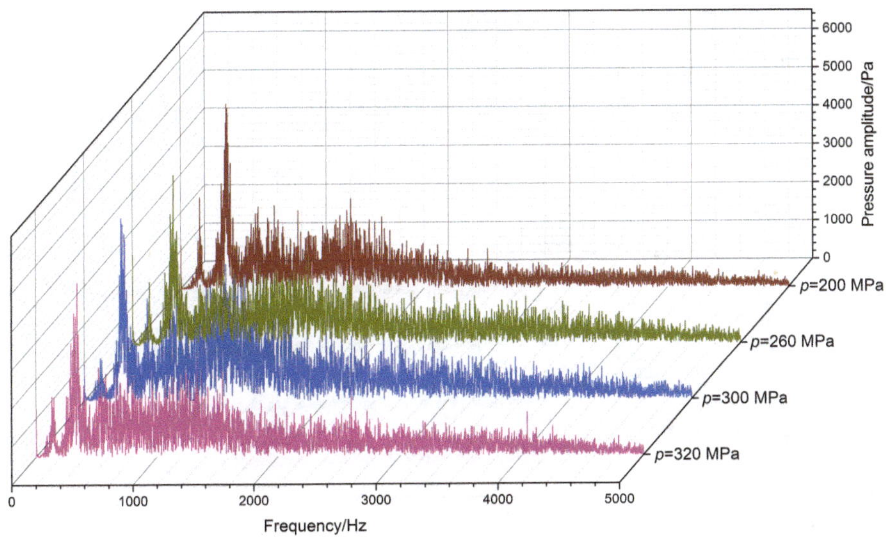

(b) $Z=Z_0+30$ mm

Fig. 3.22 Pressure fluctuations for different vertical positions

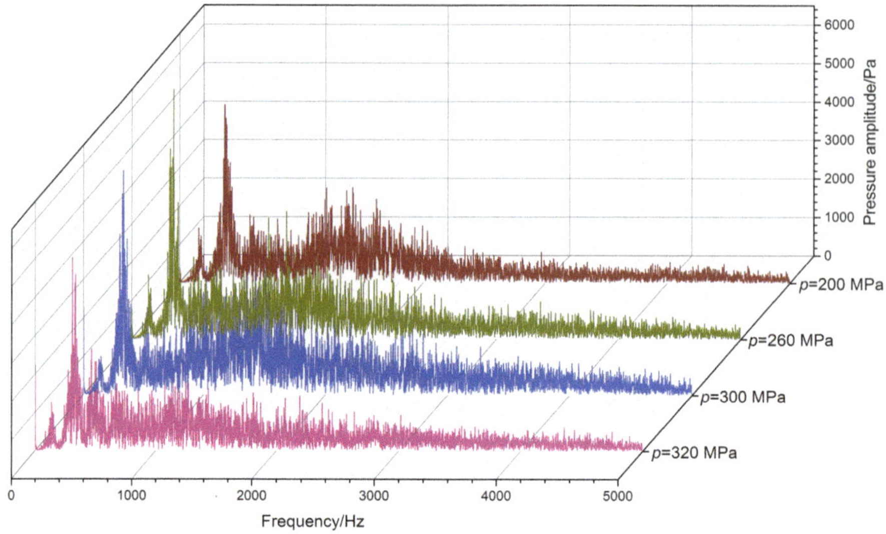

(c) Z=Z_0+60 mm

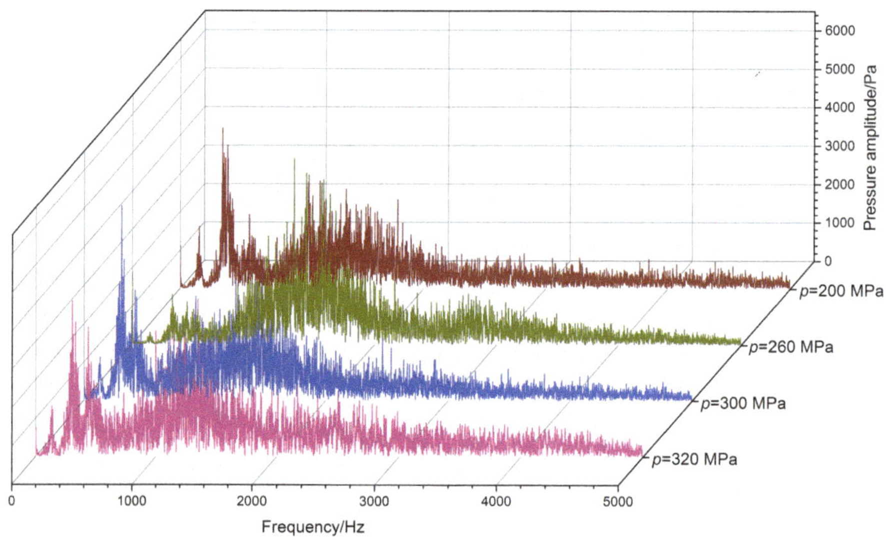

(d) Z=Z_0+105 mm

Fig. 3.22 (continued)

3.2.4 Summary

At high and ultra-high jet pressures, the impact of the submerged waterjet on the target surface is featured by dual influential effects, one originates from waterjet stagnation pressure and the other one is cavitation erosion. Based upon surface morphology characteristics, the cavitation erosion process is divided into four consecutive stages, which are accordingly represented by few small pits, sparsely distributed moderate-size pits, few large-size erosion pits and interconnected large-size erosion pits.

At the initial cavitation erosion stage, microhardness increases constantly with the jet pressure at certain impingement time. The maximum microhardness is detected within a thin layer underneath the target surface. At certain jet pressure, overall microhardness increases with the impingement time. With the development of cavitation erosion, large-size erosion pits appear and both the pit number and size show a sharp rise as cavitation erosion is reinforced.

At certain impingement time, the mass removal rate is enlarged as the jet pressure increases. Similar tendency is evident as the mass removal rate varies with the jet pressure while the standoff distance is fixed. There is an optimum standoff distance at each jet pressure regarding the mass removal rate. The influence of the standoff distance is reflected by pressure fluctuations, which exhibit two distinct peaks in frequency domain, one corresponds to the waterjet stream itself and the other is related to cavitation.

3.3 Cavitation Simulation for Submerged Waterjet and Experimental Validation

A variety of materials have been tested with waterjet [51]. As for submerged water jet, the participation of cavitation bubbles greatly enhances the impingement effect [52]. In this context, the submerged waterjet with jet pressures near 100 MPa has rarely been reported [53].

Various flow phenomena have been visualized in submerged waterjet. Typical phenomena such as jet breakup, formation and development of vortices at jet rim, and bubble formation have been widely recognized [54]. Weakness of available experimental techniques is evident in the presence of these flow phenomena, particularly as the jet pressure is high. As far as optical flow measurement technique is concerned, sparsely distributed bubbles and bubble clusters around the water jet stream hinder the penetration of laser into the jet stream. In some cases, it is even difficult to distinguish the waterjet stream from ambient fluid. Alternatively, the submerged waterjet can be treated using numerical simulation [55]. Nevertheless, some numerically obtained results, particularly unsteady solutions, still endure debates owing to uncertainties. Finite element method has been utilized to simulate stress wave propagation in impacted solid samples, but for submerged waterjet, the release of energy due to bubble collapse cannot be modelled with current FEM models [56]. In addition, as shown in Sect. 3.2, surface morphology features of the impinged solid surface can be used to reveal flow-rated information, but only some preliminary conclusions have been obtained hitherto [57].

In this section, submerged waterjet with jet pressures varying from 80 to 150 MPa is considered. Computational fluid dynamics technique is utilized to virtually visualize distributions of flow velocity, turbulent kinetic energy, as well as the void fraction of cavitation. Based on the parameters in numerical simulation, an impingement experiment is conducted with Ti-6Al-4V samples. Three aspects related to impinged samples, namely residual stress, surface morphology feature and surface roughness, are quantified using optical measurement apparatuses, which is similar to those practiced in Sect. 3.2. For comparison, non-submerged waterjets under similar jet pressure conditions are investigated.

3.3.1 Numerical Model and Procedure

The nozzle used here is schematically shown in Fig. 3.23. It is seen that the flow passage at the outlet part of the nozzle is divergent rather than straight. Such a design will lengthen the coherent segment of the jet stream immediately downstream of the nozzle. Another advantage of this nozzle is that it applies to high jet pressures and produces high impingement force. A systematic comparison between various nozzle outlet shapes has been recorded in [58].

The computational domain is displayed in Fig. 3.24 where a straight subdomain upstream of the nozzle is not shown for clarity. Water discharged from the nozzle enters a cylindrical subdomain filled with water and the streamwise direction is $+Z$ direction, as indicated in Fig. 3.24. The plane of $Z = 0$ mm overlaps with the nozzle outlet section. This cylindrical subdomain is sufficiently spacious to accommodate the development of the waterjet in both streamwise and lateral directions. Meanwhile, the streamwise length of this subdomain exceeds $33d$, where d is the diameter of the nozzle outlet section. This ensures a fully developed turbulent waterjet. Structured grids are used to discretize the entire computational

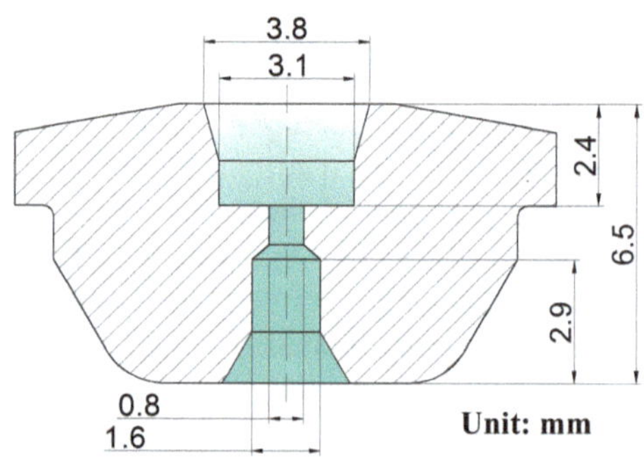

Fig. 3.23 Cross-sectional diagram of the nozzle

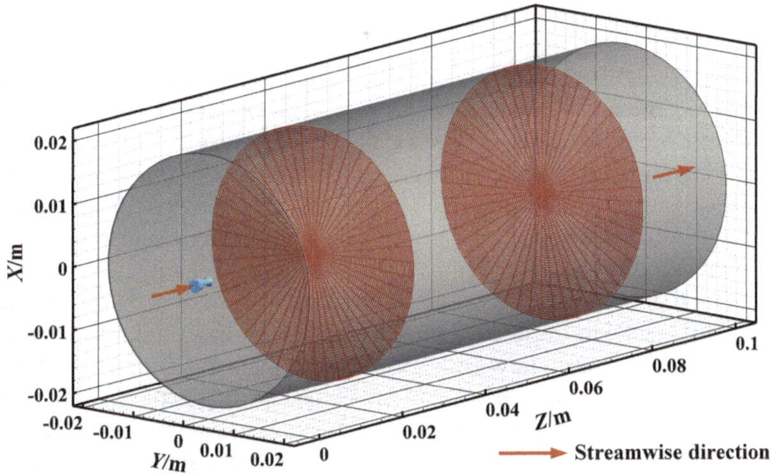

Fig. 3.24 Geometrical models of the computational domain

domain and grid refinement is executed for near-wall flow regions. With a grid independence examination, the grid scheme with total grid number of 3376780 is selected.

The renormalization group (RNG) k–ε turbulence model, as well as the Raleigh-Plesset based cavitation model is used here [59]. Numerical simulation is performed with the commercial CFD code ANSYS-CFX, which is flexible in the treatment of complex turbulent flows. Central finite difference scheme is adopted to treat the advection terms. Discretization of momentum and turbulent kinetic energy equations is accomplished using the second-order upwind scheme. Velocity inlet boundary condition is defined at the inlet of the whole computational domain. The velocity magnitude is calculated through the empirical relation between velocity, density and jet pressure. Static pressure conditions are employed at the outlet of the whole computational domain. No-slip condition is applied for all solid boundaries. Near-wall flow regions are treated with scalable wall functions.

3.3.2 Discussion of Numerical Results

Both submerged and non-submerged water jets are simulated with the same set of geometrical models. Three jet pressures, 80, 100 and 120 MPa, are adopted. Submerged waterjet and non-submerged waterjet are substantially different in terms of the environmental disturbance [60]. Numerically obtained flow parameters can lend their support to explain the difference [61].

3.3.2.1 Velocity Distributions

Velocity magnitude is representative of the impingement capability of the waterjet. For the three jet pressures considered, cross-sectional velocity distributions at $Z = 5$, 10 and 15 mm are extracted from numerically obtained results. These three standoff distances are commonly used in jet impingement practice. Given that $Z/d > 30$, the impact pressure exerted on the target sample will show obvious fluctuations [21]. Here, the selection of the standoff distance avoids this adverse situation.

For submerged waterjets, resultant cross-sectional velocity distributions are displayed in Fig. 3.25. The same velocity scale is used for all distributions in Fig. 3.25. It is seen that overall velocity magnitude increases with the jet pressure, which is especially outstanding as the jet pressure increases from 80 to 100 MPa. Meanwhile, velocity magnitudes in jet core regions are distinctly high. For each jet pressure, global velocity declines along the streamwise direction, and meanwhile, velocity in the jet core region decays rapidly. The resistance and disturbance

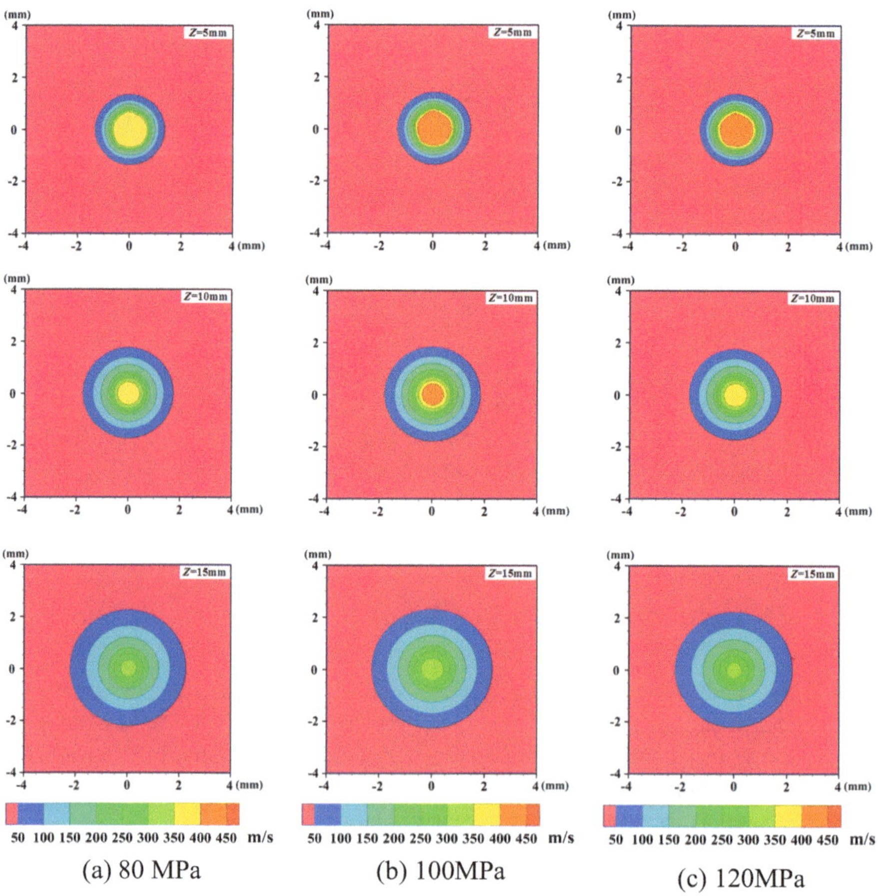

Fig. 3.25 Cross-sectional velocity distributions in the submerged waterjet stream

originating from ambient water impair the integrity of the waterjet stream. As Z increases, contour lines keep stable circular shape irrespective of the variation in the jet pressure. With the jet pressures and standoff distances considered, jet breakup and the loss of stability do not occur during the progression of the submerged waterjet.

Velocity distributions in the non-submerged water jets are exhibited in Fig. 3.26. Regarding the variation of overall velocity magnitude, the tendency indicated in Fig. 3.26 is identical with that implied in Fig. 3.25. Nevertheless, radial diffusion of high-velocity area is less severe in Fig. 3.26. Another characteristic of Fig. 3.26 is that for the same jet pressure, from $Z = 5$ to 10 mm, velocity decay along the jet direction is much weak compared to that shown in Fig. 3.25. Such a tendency helps to validate the suitability of this range of standoff distance in practical applications of jet impingement. In this context, velocity will attenuate dramatically as the standoff distance exceeds a threshold value, which has been proved in [62].

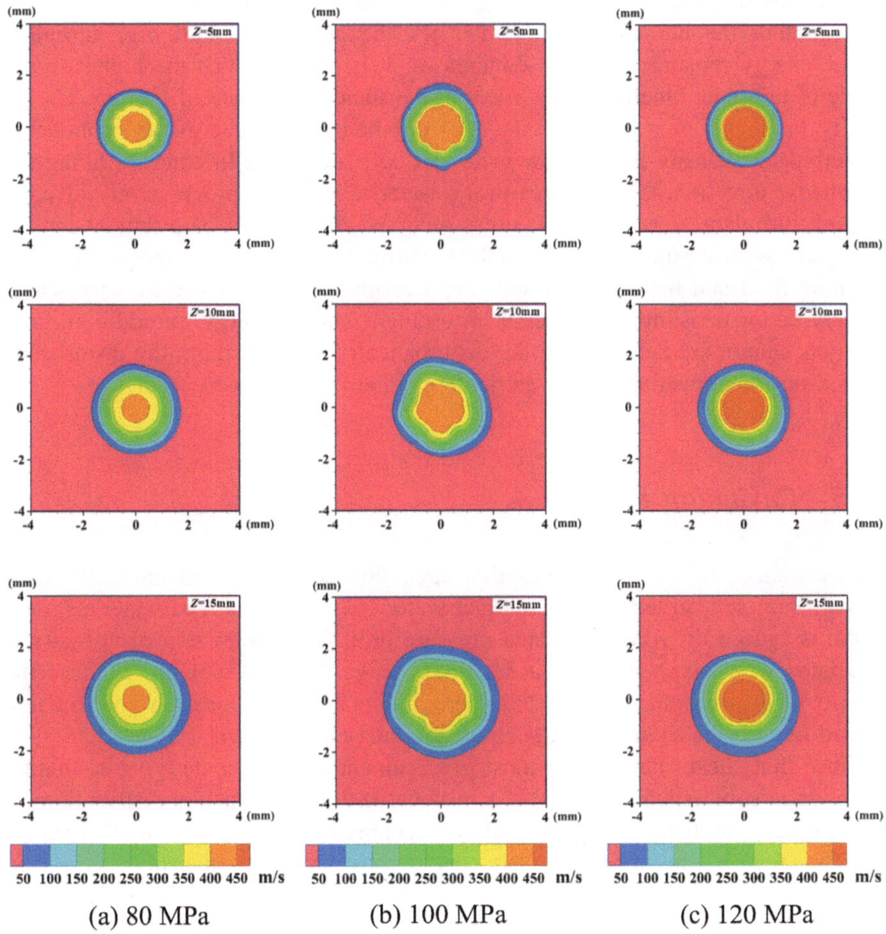

Fig. 3.26 Cross-sectional velocity distributions for non-submerged waterjet

Furthermore, most velocity contour lines displayed in Fig. 3.26 cannot maintain circular shape, as differs clearly from that depicted in Fig. 3.25. For non-submerged waterjet, the difference of density between working medium and ambient fluid contributes significantly to the wavy structures at the jet stream edge [63].

3.3.2.2 Turbulent Characteristics

Apart from flow quantities expressed in an averaged manner, turbulent features are of particular interest at high jet pressures. Velocity fluctuations are in direct association with the impingement effect on target surfaces. Previous studies have confirmed that drastic velocity fluctuations lead to irregular footprints on target surface. Turbulent kinetic energy is inherently a second-order momentum of velocity fluctuation and can be obtained with the turbulence model used in the present simulation. In Fig. 3.27, a comparison is illustrated through cross-sectional distributions of turbulent kinetic energy at jet pressure of 100 MPa.

Regarding the non-submerged water jet, the annular area of high turbulent kinetic energy remains stable as Z increases. It is observed that the highest magnitude of turbulent kinetic energy arises at the standoff distance of 10 mm. The jet core is dominated by a rather low level of turbulent kinetic energy, as implies that the turbulent intensity is very low in the jet core area [64]. In contrast, along the streamwise direction, the core area in the waterjet stream is progressively invaded by high turbulent kinetic energy; cross-sectional distribution of turbulent kinetic energy tends to be uniform. Shear effect, nurtured both between the jet and surrounding fluid and between adjacent layers residing in the jet stream, serves as a primary factor boosting the increase in turbulent kinetic energy. In addition, both waterjets cannot are subjected to the disturbance of ambient fluid; the disturbance exerted on the non-submerged waterjet is evidently more severe.

3.3.3 Cavitation Prediction

In this section, it is unambiguous that cavitation is the most distinct difference between non-submerged and submerged waterjets. For submerged waterjets, cavitation is caused by vorticity-related pressure drop. The production of vorticity is associated with the shear effect at jet stream edge. Cavitation might enhance the severity of pressure fluctuations [65]. Here, the quantity of cavitation void fraction is used to describe cavitation in submerged waterjet, as shown in Fig. 3.28.

Since that shear effect is the most predominant impetus underlying cavitation inception in submerged waterjets, it is reasonable that void fraction distributions at the same standoff distance but with different jet pressures are similar. As Z increases, cavitation region tends to be expanded; concurrently, cavitation intensity is improved consistently. Such a tendency hinges upon the laterally diffused jet stream with adequate kinetic energy to sustain the shear effect. Further downstream, there is a

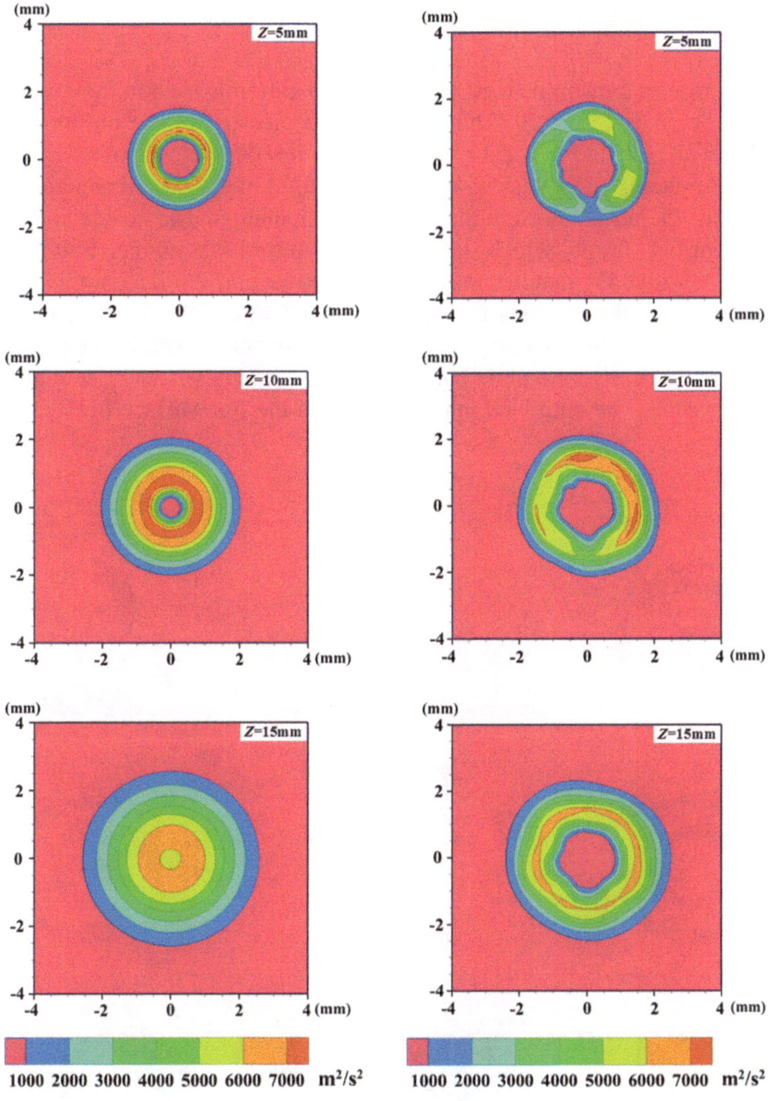

(a) Submerged waterjet at jet pressure of 100 MPa

(b) Non-submerged waterjet at jet pressure of 100 MPa

Fig. 3.27 Cross-sectional distributions of turbulent kinetic energy

critical position where cavitation vanishes thoroughly. As far as cavitation erosion is concerned, it can be predicted from Fig. 3.28 that surface damage due to cavitation will show annular shape. Similar cavity shape is found in Soyama's work in which cavitation is produced through an annular nozzle which simultaneously injects two streams of liquids with different initial velocities into air [66].

3.3.4 Experimental Setup

A waterjet machine manufactured by Flow Corporation is used in the experimental study, as shown in Fig. 3.29. The maximum jet pressure of the machine exceeds 380 MPa. Traverse and vertical motions of the nozzle are controlled by an automatic control facility. Dimensions of the nozzle used in the experiment are identical with those of the nozzle used in the numerical simulation. The nozzle is embedded in a supporting frame which is rigidly connected to its neighboring parts. Cylindrical Ti-6Al-4V samples, 40 mm in diameter and 2.5 mm in thickness, are used as target samples. Jet pressures of 80, 100, 120 and 150 MPa are predefined in the experiment. During the experiment, samples are submerged in the tank located at the bottom of the whole unit. The depth of the samples is large enough so that ambient air cannot be entrained into the tank via the free surface [67].

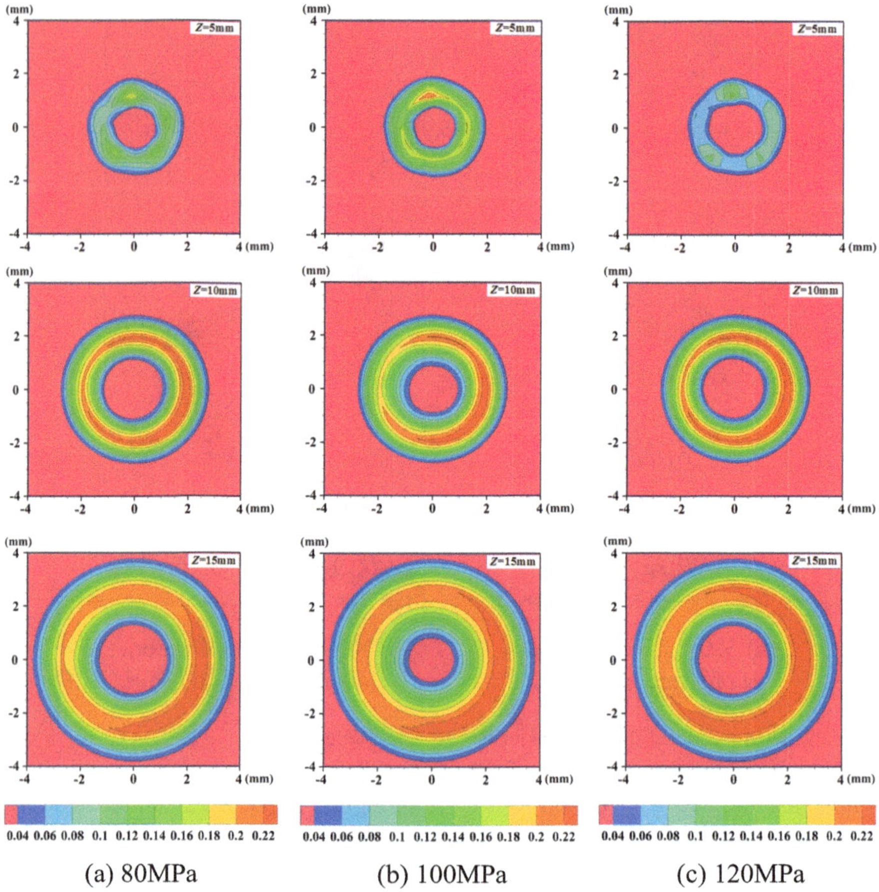

Fig. 3.28 Void fraction of cavitation for submerged waterjet

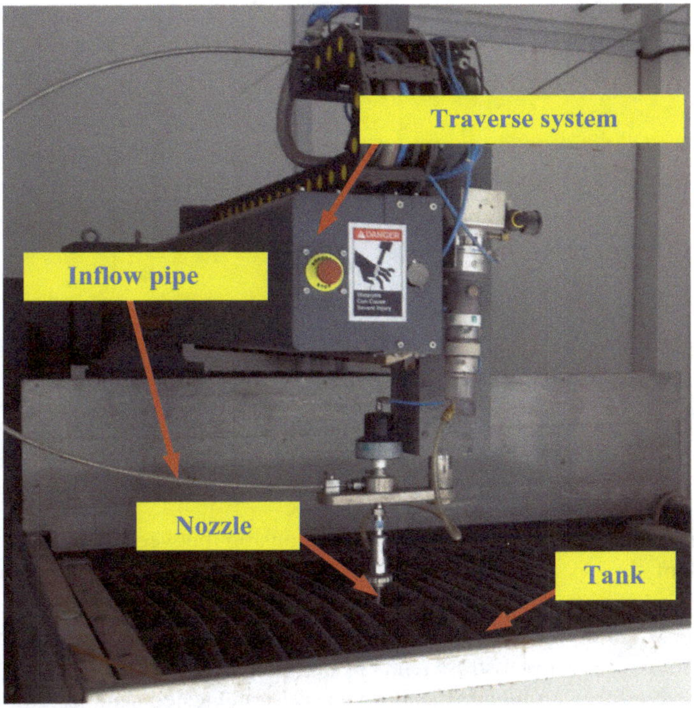

Fig. 3.29 High-pressure waterjet machine

3.3.5 Experimental Results

3.3.5.1 Residual Stress

At the jet pressure of 80 MPa and standoff distance of 15 mm, no sunken area or pits are detected on the impacted sample surface. Then 11 samples are tested to examine the variation of the residual stress with the impingement time. Hitherto, the difference of residual stress distribution between submerged and non-submerged waterjets has not been explicitly explained [68]. An X-ray stress analyzer has been used to measure residual stress for local impinged areas [69]. The present study also uses such an instrument together with oblique fixed Ψ method. Ψ is the angle between the normal vector of the impinged surface and the normal vector of the diffractive face. Values of Ψ, 0°, 24°, 35° and 45° are adopted. Stress constant is −601 MPa per degree. Both the obtained residual stress and measurement errors are plotted in Fig. 3.30.

Residual stress distribution on impinged solid surface is compatible with the distribution of the impact pressure [70]. Here, in consideration of the tiny waterjet stream, residual stress shown in Fig. 3.30 can be deemed as representative of area-averaged residual stress on the impinged area. Negative values denote

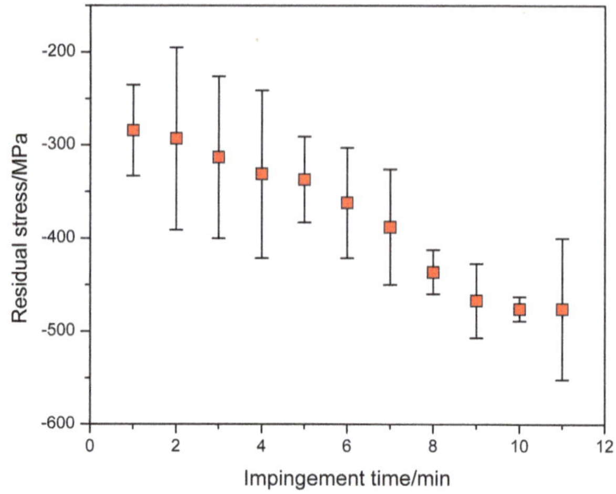

Fig. 3.30 Variation of the residual stress with the impingement time

compressive stress. It is seen that the relationship between residual stress and impingement time is nearly linear, as is in accordance with general knowledge. From the perspective of residual stress magnitude, the strengthening effect of the submerged waterjet on solid surface is remarkable. As the impingement time exceeds 9 min, the residual stress tends to be stable, as indicated in Fig. 3.30. Such a trend also conforms to the physical properties of the tested samples. Although the existence of cavitation is numerically proved under such a situation, cavitation effect is not explicitly recognized.

3.3.5.2 Cavitation Effect Identification

For the jet pressure of 100 MPa and standoff distance of 5.0 mm, two samples are impinged by submerged and non-submerged waterjets, respectively. Identical impingement time of 60 s is specified. A close examination of impinged local surface provides small-scale characteristics [71]. Thus an AxioCSM700 confocal microscope is used to capture surface morphology images of the tested samples [72]. Two amplified images corresponding to the two impinged local surfaces are exhibited in Fig. 3.31. In Fig. 3.31a, the penetration of the submerged waterjet into the sample surface is not apparent, but the annular polished area justifies the effect of cavitation. In contrast, as shown in Fig. 3.31b, an resultant hole on the sample surface is seen, and the impingement force depends on the jet velocity. Around the hole, there is also a polished annular area, which is produced due to relatively low kinetic energy. In addition, the perimeter of the hole shown in Fig. 3.31b is not circular, which is caused by wavy structures at the interface between the waterjet stream and surrounding air.

Then another sample is impinged by submerged waterjet at the jet pressure of 100 MPa and a duration period of 60 s. The confocal microscope is used to

Fig. 3.31 Local surfaces impinged by waterjet at 100 MPa and duration time of 60 s

(a) Local area impinged by submerged waterjet

(b) Local area impinged by non-submerged waterjet

measure surface roughness on the impinged surface. Two lines emanating radially from the impingement center are monitored and distributions of local heights along these two lines are plotted in Fig. 3.32. The two distribution curves share the trait that a relatively smooth area is situated between two coarse areas. Such a situation is in accordance with the results shown in Fig. 3.31 and reconfirms the cavitation effect which is concentrated on an annular area.

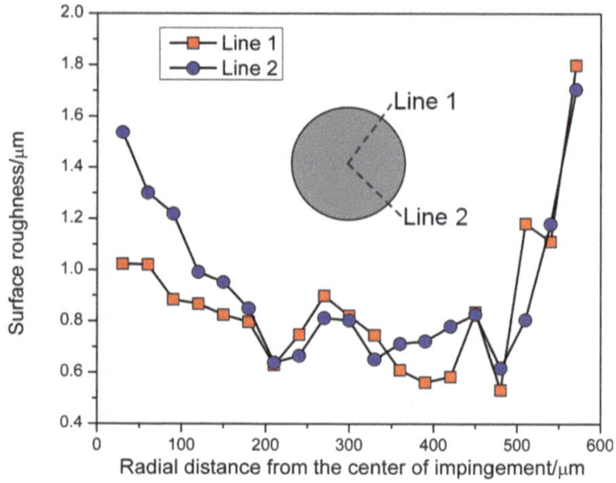

Fig. 3.32 Surface roughness on the surface impinged by submerged waterjet at 100 MPa

3.3.5.3 Surface Morphological Features

For the jet pressure of 120 MPa, the duration of the jet impingement is specified as 3 min. Obtained surface morphology features of impinged samples are depicted in Fig. 3.33. According to Fig. 3.28c, cavitation is not influential in submerged waterjet at the standoff distance of 5 mm, thus the impingement effect depends primarily on jet kinetic energy. Profile of the resultant pit on the impacted surface shown in Fig. 3.33a also supports this conjecture. As the standoff distance is shifted to 10 mm, as shown in Fig. 3.33b, the shape of the bottom of the pit serves as an indicator of cavitation effect. More specifically, the sunken part at the bottom of the pit is related to cavitation effect. Moreover, the perimeters of the pits shown in Figs. 3.33a, b are perfectly circular.

With respect to the impingement effect of the non-submerged waterjet, three standoff distances are selected and the surface morphology features are illustrated in Figs. 3.33c, d, e, respectively. In both Figs. 3.33c, d, waterjet evidently penetrates into the sample surface. The irregularity of the perimeter of the hole is apparent, particularly at the standoff distance of 15 mm. With a further increase of 5 mm in the standoff distance, as shown in Fig. 3.33e, there is a relatively shallow footprint left on the impinged sample surface. Meanwhile, the perimeter of the resultant pit left on the surface is coarse as well. According to general viewpoints, liquid droplets contained in non-submerged waterjet results in the formation of small-scale cracks on the impinged sample surface, while for submerged waterjets, bubble collapse is a crucial factor underlying the damage to the target surface [73]. For the latter, an effective instrument for evaluating the cavitation damage is still badly expected.

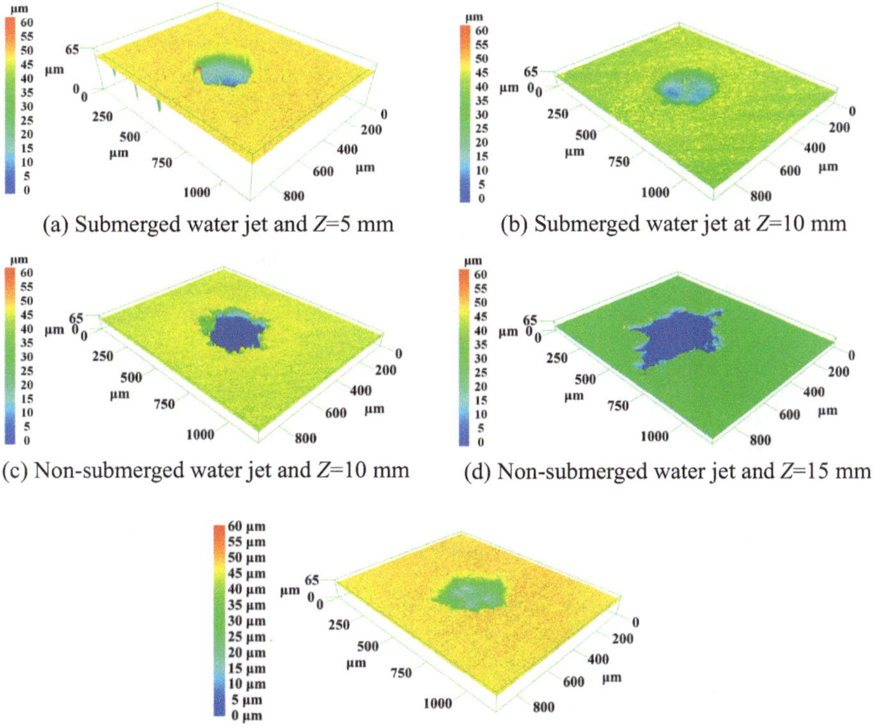

(a) Submerged water jet and Z=5 mm

(b) Submerged water jet at Z=10 mm

(c) Non-submerged water jet and Z=10 mm

(d) Non-submerged water jet and Z=15 mm

(e) Non-submerged water jet and Z=20 mm

Fig. 3.33 Surface morphology of impinged surfaces

3.4 Concluding Remarks

Numerical simulation enables the examination of flow parameter distributions in tiny water jet stream, as well as the development of cavitation phenomenon. For jet pressures considered, overall jet kinetic energy attenuates rapidly in streamwise direction. Concurrently, the influence of ambient water on jet core area is consistently intensified. Cross-sectional distributions of cavitation void fraction indicate that the existence of annular cavity is in agreement with the formation mechanism dominated by shear effect.

Experiments of submerged waterjet impingement with Ti-6Al-4V samples not just provide evidence for cavitation damage but also yield quantitative information of surface morphology features. Residual stress increases nearly linearly with impingement time. Both the shape and surface roughness associated with the footprints on impinged surfaces are in good agreement with jet flow characteristics.

At high jet pressures, contributions of cavitation and jet pressure should be jointly considered. This implies that both factors are capable of damaging target surface. In this context, the debate stems from the question that which factor is more influential for the damage to impinged sample surface. It is certain that as the jet pressure reaches ultra-high level, jet kinetic energy is definitely the leading factor, and a single hole, instead of multiple pits will represent the resultant footprints.

References

1. Dalfré Filho JG, Assis MP, Genovez AIB. Bacterial inactivation in artificially and naturally contaminated water using a cavitating jet apparatus. J Hydro Environ Res. 2015;9:259–267.
2. Shaffer F, Savaş Ö, Lee K, de Vera G. Determining the discharge rate from a submerged oil leak jet using ROV video. Flow Meas Instrum. 2015;43:34–46.
3. Wright MM, Epps B, Dropkin A, Truscott TT. Cavitation of a submerged jet. Exp Fluids. 2013;54:1541.
4. Chahine GL, Kapahi A, Choi JK, Hsiao CT. Modeling of surface cleaning by cavitation bubble dynamics and collapse. Ultrason Sonochem. 2015;29:528–549.
5. Liu H, Shao Q, Kang C, Gong C. Impingement capability of high-pressure submerged water jet: numerical prediction and experimental verification. J Cent South Univ. 2015;22:3712–3721.
6. Ganippa L, Bark G, Andersson S, Chomiak J. Cavitation: a contributory factor in the transition from symmetric to asymmetric jets in cross-flow nozzles. Exp. Fluids. 2004;36:627–634.
7. Wilms J, Beck T, Sorensen CM, Hosni MH, Eckels SJ, Tomasi D. Experimental measurements and flow visualization of water cavitation through a nozzle. In: Proceedings of the ASME 2014 international mechanical engineering congress and exposition IMECE2014; Montreal, Quebec, Canada; 2014.
8. Cheng Y, del Mar Torregrosa M, Villegas A, Diez FJ. Time resolved scanning PIV measurements at fine scales in a turbulent jet. Int J Heat Fluid Flow. 2011;32:708–718.
9. Gopalan S, Katz J, Knio O. The flow structure in the near field of jets and its effect on cavitation inception. J Fluid Mech. 1999;398:1–43.
10. Nobukazu S, Hideki H, Yukio I, Masaru S, Kozo S. Analysis of submerged water jets by visualization method flow pattern and self-induced vibration of jet. J Vis. 2004;7:281–289.
11. Nobel AJ, Talmon AM. Measurements of the stagnation pressure in the center of a cavitating jet. Exp Fluids. 2012;52:403–415.
12. Zeleňák M, Foldyna J, Scucka J, Hloch S, Riha Z. Visualisation and measurement of high-speed pulsating and continuous water jets. Measurement. 2015;72:1–8.
13. Geller S, Uphoff S, Krafczyk M. Turbulent jet computations based on MRT and Cascaded Lattice Boltzmann models. Comput Math Appl. 2012;65(12):1956–1966.
14. Kendil FZ, Danciu DV, Schmidtke M, Salah AB, Lucas D, Krepper E, Mataoui A. Flow field assessment under a plunging liquid jet. Prog Nucl Energy. 2012;56:100–110.
15. Wang Z, He P, Lv Y, Zhou J, Fan J, Cen K. Direct numerical simulation of subsonic round turbulent jet. Flow Turbul Combust. 2010;84(4):669–686.
16. Lawson NJ, Davidson MR. Self-sustained oscillation of a submerged jet in a thin rectangular cavity. J Fluids Struct. 2001;15:59–81.
17. Ghasemi A, Roussinova VT, Balachandar R, Barron R. Reynolds number effects in the near-field of a turbulent square jet. Exp Thermal Fluid Sci. 2015;61:249–258.
18. Shim YM, Sharma RN, Richards P. Proper orthogonal decomposition analysis of the flow field in a plane jet. Exp Thermal Fluid Sci. 2013;51:37–55.
19. Zhou Y, Lucey AD. Fluid-structure-sound interactions and control. Lecture notes in mechanical engineering. Berlin: Springer; 2016;229–234.
20. Fitzgerald JA, Garimella SV. A study of the flow field of a confined and submerged impinging jet. Int J Heat Mass Transf. 1998;41:1025–1034.

21. Soyama H, Yanauchi Y, Sato K, Ikohagi T, Oba R, Oshima R. High-speed observation of ultrahigh-speed submerged water jets. Exp Thermal Fluid Sci. 1996;12:411–416.
22. Zhang F, Liu H, Junchao X, Tang C. Experimental investigation on noise of cavitation nozzle and its chaotic behaviour. Chin J Mech Eng. 2013;26(4):758–762.
23. Peng G, Shimizu S. Progress in numerical simulation of cavitating water jets. J Hydrodyn. 2013;25:502–509.
24. Hutli EAF, Nedeljkovic MS. Frequency in shedding/discharging cavitation clouds determined by visualization of a submerged cavitating jet. J Fluids Eng Trans ASME. 2008;130:021304-1-8.
25. Sato K, Taguchi Y, Hayashi S. High speed observation of periodic cavity behavior in a convergent-divergent nozzle for cavitating water jet. J Flow Control Meas Vis. 2013;1:102–107.
26. Hongxiang H, Zheng Y, Qin CP. Comparison of Inconel 625 and Inconel 600 in resistance to cavitation erosion and jet impingement erosion. Nucl Eng Des. 2010;240:2721–2730.
27. Wang Y, Qiu L, Reitz RD, Diwakar R. Simulating cavitating liquid jets using a compressible and equilibrium two-phase flow solver. Int J Multiph Flow. 2014;63:52–67.
28. Peters A, Sagar H, Lantermann U, el Moctar O. Numerical modelling and prediction of cavitation erosion. Wear. 2015;338–339:189–201.
29. Li Z, Pourquie M, van Terwisga T. Assessment of cavitation erosion with a URANS method. J Fluids Eng Trans ASME. 2014;136:041101-1-11.
30. Kang C, Liu H, Li X, Zhou Ya, Cheng X. A numerical and experimental study of oblique impact of ultra-high pressure abrasive water jet. Adv Mech Eng. 2016;8(3):1–14.
31. Verhaagen B, Rivas DF. Measuring cavitation and its cleaning effect. Ultrason Sonochem. 2015;29:619–628.
32. Petkovšek M, Dular M. Simultaneous observation of cavitation structures and cavitation erosion. Wear. 2013;300:55–64.
33. Anantharamaiah N, Tafreshi H, Pourdeyhimi B. Numerical simulation of the formation of constricted waterjets in hydroentangling nozzles: effects of nozzle geometry. Chem Eng Res Des. 2006;84(3):231–238.
34. Alturki F, Abouel-Kasem A, Ahmed SM. Characteristics of cavitation erosion using image processing techniques. J Tribol Trans ASME. 2013;135:014502-1-7.
35. ASTM G134-95(2010)e1. Standard test method for erosion of solid materials by a cavitating liquid jet; 2010.
36. Hutli EAF, Nedeljkovic MS, Radovic NA. Mechanics of submerged jet cavitating action: material properties, exposure time and temperature effects on erosion. Arch Appl Mech. 2008;78:329–341.
37. Hu H, Jiang S, Tao Y, Xiong T, Zheng Y. Cavitation erosion and jet impingement erosion mechanism of cold sprayed Ni–Al_2O_3 coating. Nucl Eng Des. 2011;241:4929–37.
38. Kang C, Liu H, Li X, Cheng X. Cavitation in submerged water jet at high jet pressure. In: Proceedings of the ASME/JSME/KSME joint fluids engineering conference, Seoul, South Korea; 2015 July 26–31.
39. Sahaya Grinspan A, Gnanamoorthy R. Development of a novel oil cavitation jet peening system and cavitation jet erosion in Aluminum alloy, AA 6063-T6. J Fluids Eng Trans ASME. 2009;131:061301-1-8.
40. Fortes Patella R, Challier G, Reboud J-L, Archer A. Energy balance in cavitation erosion: from bubble collapse to indentation of material surface. J Fluids Eng Trans ASME. 2013;135:011303-1-11.
41. Abouel-Kasem A, Ezz El-Deen A, Emara KM, Ahmed SM. Investigation into cavitation erosion pits. J Tribol Trans ASME. 2009;131:031605-1-7.
42. Choi J-K, Jayaprakash A, Kapahi A, Hsiao C-T, Chahine GL. Relationship between space and time characteristics of cavitation impact pressures and resulting pits in materials. J Mater Sci. 2014;49:3034–3051.
43. Soyama H, Takeo F. Comparison between cavitation peening and shot peening for extending the fatigue life of a duralumin plate with a hole. J Mater Process Technol. 2016;227:80–87.

44. Osterman A, Bachert B, Sirok B, Dular M. Time dependant measurements of cavitation damage. Wear. 2009;266:945–951.
45. Sun Z, Kang X, Wang X. Experimental system of cavitation erosion with water-jet. Mater Des. 2005;26:59–63.
46. Choi J-K, Jayaprakash A, Chahine GL. Scaling of cavitation erosion progression with cavitation intensity and cavitation source. Wear. 2012;278–279:53–61.
47. Vickers GW, Houlston R. Modelling the erosion efficiency of cavitating cleaning jets. Appl Sci Res. 1983;40:377–391.
48. Franc J-P, Riondet M, Karimi A, Chahine GL. Material and velocity effects on cavitation erosion pitting. Wear. 2012;274–275:248–259.
49. Li J, Yi M, Shen Z, Ma S, Zhang X, Xing Y. Experimental study on a designed jet cavitation device for producing two-dimensional nanosheets. Sci China Technol Sci. 2012;55:2815–2819.
50. De Giorgi MG, Ficarella A, Tarantino M. Evaluating cavitation regimes in an internal orifice at different temperatures using frequency analysis and visualization. Int J Heat Fluid Flow. 2013;39:160–172.
51. Venkatesh VC. Machining of glass by impact processes. J Mech Working Technol. 1983;8:247–260.
52. Tönshoff HK, Kroos F, Marzenell C. High-pressure water peening-a new mechanical surface-strengthening process. Ann CIRP. 1997;46(1):113–116.
53. Kang C, Zhou L, Yang M, Wang Y. Experiment study on cavitating waterjet induced by a central body in the nozzle. J Eng Thermophys. 2013;34(12):2275–2278.
54. Lemanov VV, Terekhov VI, Sharov KA, Shumeiko AA. An experimental study of submerged jets at low Reynolds numbers. Tech Phys Lett. 2013;39(5):421–423.
55. Guo B, Langrish TAG, Fletcher DF. An assessment of turbulence models applied to the simulation of a two-dimensional submerged jet. Appl Math Model. 2001;25:635–653.
56. Anwar S, Axinte DA, Becker AA. Finite element modelling of abrasive waterjet milled footprints. J Mater Process Technol. 2013;213:180–193.
57. Ayed Y, Germain G, Ammar A, Furet B. Degradation modes and tool wear mechanisms in finish and rough machining of Ti17 Titanium alloy under high-pressure water jet assistance. Wear. 2013;305:228–237.
58. Hitoshi S. Effect of nozzle geometry on a standard cavitation erosion test using a cavitating jet. Wear. 2013;297:895–902.
59. Bakir F, Rey R, Gerber AG, Belamri T, Hutchinson B. Numerical and experimental investigations of the cavitating behavior of an inducer. Int J Rotating Mach. 2004;10:15–25.
60. Kang C, Liu H. Turbulent features in the coherent central region of a plane water jet issuing into quiescent air. J Fluids Eng. 2014;136(8):081205.
61. Kotsovinos N, Angelidis P. The momentum flux in turbulent submerged jets. J Fluid Mech. 1991;229:453–470.
62. Babarsad MS, Jahromi HM, Kashkooli H, Samani HMV, Sedghi H. Experimental study of maximum velocity and effective length in submerged jet. Indian J Sci Technol. 2013;6(1): 18–20.
63. Weiland C, Vlachos PP. Round gas jets submerged in water. Int J Multiph Flow. 2013;48: 46–57.
64. Rosler RS, Bankoff SG. Large-scale turbulence characteristics of a submerged water jet. AIChE J. 1963;9(5):672–676.
65. Franklin RE, Mcmillan J. Noise generation in cavitating flows, the submerged jet. J Fluids Eng. 1984;106:336–341.
66. Soyama H, Kikuchi T, Nishikawa M, Takakuwa O. Introduction of compressive residual stress into stainless steel by employing a cavitating jet in air. Surf Coat Technol. 2011;205:3167–3174.
67. Qu X, Goharzadeh A, Khezzar L, Molki A. Experimental characterization of air-entrainment in a plunging jet. Exp Thermal Fluid Sci. 2013;44:51–61.
68. Rajesh NR, Sundararaghavan V, Babu NR. A novel approach for modelling of water jet peening. Int J Mach Tools Manuf. 2004;44:855–863.

69. Odhiambo D, Soyama H. Cavitation shotless peening for improvement of fatigue strength of carbonized steel. Int J Fatigue. 2003;25:1217–1222.
70. Qin M, Dongying J, Oba R. Investigation of the influence of incidence angle on the process capability of water cavitation peening. Surf Coat Technol. 2006;201:1409–1413.
71. Takakuwa O, Soyama H. The effect of scanning pitch of nozzle for a cavitating jet during overlapping peening treatment. Surf Coat Technol. 2012;206:4756–4762.
72. Kang C, Liu H. Small-scale morphological features on a solid surface processed by high-pressure abrasive water jet. Materials. 2013;6(8):3514–3529.
73. Sahaya Grinspan A, Gnanamoorthy R. Surface modification by oil jet peening in Al alloys, AA6063-T6 and AA6061-T4, part 2: surface morphology, erosion, and mass loss. Appl Surf Sci. 2006;253:997–1005.

Chapter 4
Motion of Bubble

Abstract The bubbly flow is an important flow pattern in fluids engineering. A variety of bubbly flows have been witnessed so far; bubble size and bubble volume fraction are two factors that affect the appearance of the bubbly flow. In consideration of the shortage of numerical simulation in the presence of bubbly flows, the supplement of the knowledge of bubbly flows depends heavily on the advancement and application of measurement techniques. Another point associated with bubbly flows lies in that many conclusions obtained are associated with certain experiment rig or specific engineering application and are therefore difficult to generalize. In this section, two cases relevant to the bubble trapped in liquid are presented. One is the rising bubble in stationary water and the other is the bubble released in horizontal water flow. Experimental techniques, in conjunction with developed image-processing code, are used. Geometric and motion quantities of bubble are adopted to describe the bubble characteristics.

4.1 Rising Bubble in Stationary Water

The enthusiasm for bubbly flow has never faded during the exploration of fluid dynamics and multiphase flows. Bubble size and its spatial distribution are two key factors that determine the motion and dynamics characteristics of bubbly flows [1, 2]. Fundamentally, the understanding of the movement of even a single bubble in bubbly flows is inadequate. In this context, multiple factors such as operation condition, liquid property and bubble production strategy contribute to the diversity of bubble trajectory, bubble geometry and bubble velocity [3, 4].

The application of the CFD technique facilitates the identification and description of various bubbly flow structures [5]. Nevertheless, the validation of numerical algorithms and grid scheme still suffers from arguments, in particular for the simulation of the correlation between bubble parameters and turbulent fluctuations.

Bubbly flow research has benefited from the development of flow measurement and visualization techniques. Of significance is the extensive practice using most

C. Kang et al., *Methods for Solving Complex Problems in Fluids Engineering*,
https://doi.org/10.1007/978-981-13-2649-3_4

advanced techniques such as high-speed photography (HSP), laser Doppler velocimetry (LDV), particle image velocimetry (PIV), planar laser induced fluorescence (PLIF) [6–9]. Regarding the identification of bubbles in captured images, techniques such as glare point velocimetry and sizing (GPVS) and planar fluorescence for bubble imaging (PFBI) offer considerable assistance [10, 11]. Thereby, the effect of nozzle diameter on bubbles turns to be traceable. In [12], it is argued that small nozzle diameter leads to ellipsoidal bubble and spiral bubble rising trajectory, while large nozzle diameter is associated with spherical bubble and zigzag bubble trajectory. It is confirmed that bubble shape, trajectory and terminal velocity are interrelated and bear clear nozzle effect [13]. Brücker et al. find that bubble trajectory shifts from a straight line to a spiral route as the bubble travels downstream and away from the nozzle [14]. Although efforts have been devoted to the investigation of nozzle effect, quantitative and generalizable conclusions are still greatly expected. In particular, it is desirable to validate established conclusions through various cases and statistical results.

Bubble velocity characteristics serve as an unambiguous indication of the state of bubble evolution. Bubble velocity is related to multiple factors such as liquid property, liquid contamination, bubble size and air injection mode [15, 16]. Maldonado et al. focused on the influence of bubble size on bubble rising velocity in stationary water [17]. In [18], Ortiz-Villafuerte et al. concluded that the contamination of water causes the decline of bubble terminal velocity. Bubble terminal velocity in liquids containing inorganic salt is investigated in [19], and it is proved that as the inorganic salt concentration improves, bubble oscillation attenuates, and bubble terminal velocity decreases. For CO_2 bubbles in static ionic liquids, it is demonstrated that bubble velocity in a state of equilibrium varies inversely with equivalent bubble diameter [20]. The work of Okawa et al. deserves much attention, they argue that the initial stage of bubble evolution has profound effect on bubble velocity and oscillation [21]. Until now, the demarcation of bubble development stages is not clear, and the main reason lies in the absence of executable judgment method.

In this section, the effect of initial bubble diameter on the evolution of bubble shape and bubble dynamics characteristics is to be described. An experiment rig is built to generate rising bubbles in stationary water through injecting air into a transparent water tank. Nozzles with different outlet diameters are utilized to produce bubbles of different size. Bubble images are acquired instantaneously using high-speed photography. An image-processing code is specially developed to identify bubble profile in the images and to calculate bubble geometrical parameters and bubble velocity. Furthermore, with the code, the extraction and verification of bubble characteristics from statistical aspects are enabled. A comparison is implemented between bubbles with different initial size. The analysis covers primary factors underlying the variety of bubble shape, bubble movement trajectory and bubble velocity. Through adjusting the flow rate of injected air, the acquired bubble data encompass a wide range of bubble size. Then a statistical investigation is performed to correlate bubble terminal velocity and bubble size. Moreover, a statistical relationship between the bubble aspect ratio and Weber number is anticipated to be established.

4.1.1 Experimental Set-Up and Image-Processing Code

4.1.1.1 Experimental Rig

The configuration of major components of the experimental system is schematically shown in Fig. 4.1. A rectangular water tank made of plexiglass is used to accommodate pure water. The tank has identical cross sections with dimensions of 150 mm × 150 mm along vertical direction, and the height of the water tank is 500 mm. At the symmetric center of the bottom wall of the tank, a nozzle is installed vertically via a ventilation hole. The nozzle is connected to a syringe pump through a plastic tube, and a check valve is used to avoid the backflow of water from the tank to the pump. Another important function of the valve is to change the flow rate of the injected air, q_V, and then to change the initial bubble size. The generation frequency of the bubble is regulated through adjusting operation parameters of the syringe pump. During the experiment, both water temperature and the injected air temperature are kept at 14.0 °C, and the water level is kept at 420 mm relative to the bottom wall of the water tank. Six nozzles with the outlet diameter D_0 of 3.00, 2.20, 1.60, 0.60, 0.19, and 0.16 mm, are used in the experiment.

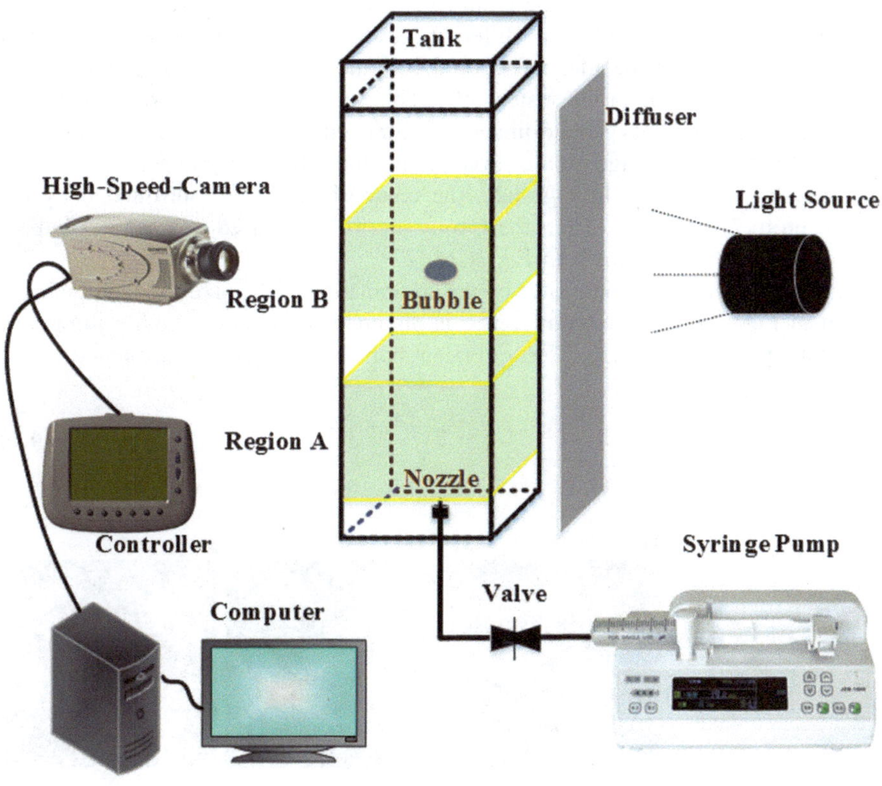

Fig. 4.1 Schematic view of the bubble visualization experiment system

Bubble images are recorded using an OLYMPUS I-SPEED 3 high-speed camera, as shown in Fig. 4.1. A light source of OLYMPUS ILP-2 is employed to assist the operation of the camera. The light passes through an acrylic diffuser plate of 5 mm in thickness before reaching the water tank. The acrylic diffuser plate plays the function of uniformizing the incident light over the depth of view (DOV) plane and thereby the error due to light diffusion is suppressed. In consideration of the magnitude of bubble rising velocity, the exposure time of the camera is set to 1 ms.

The rising bubble will attain a state of equilibrium after traveling a vertical distance [22]. As for such a distance, a suggestion of 75 mm is presented [23]. In this connection, a rectangular coordinates system is defined. The origin of the coordinates system overlaps with the center of the nozzle outlet section. The *x*-*y* plane is in parallel with the diffuser plate and *y* points upwards. Two regions, Region A with *y* ranging from 0 to 112 mm, and Region B with *y* ranging from 200 to 290 mm are monitored to seek bubble characteristics at different states and to record bubble data for statistical analysis.

4.1.1.2 Bubble Image Processing Preparation

To identify the bubble in the image and to acquire quantitative bubble information, a code is developed based on the commercial MATLAB software. Bubble image consists of a series of pixels with different grayscale values, as serves as the most fundamental base for the idea of recognizing and plotting bubble profile in the bubble image [24]. The primary steps of bubble image processing are displayed in Fig. 4.2. Firstly, the background image is subtracted from the raw image, then the image is treated with image binarization and median filtering approaches, as shown in Fig. 4.2a, b [25, 26]. In Fig. 4.2c, the color of the cavity surrounded by the bubble profile is inverted. Then the Canny algorithm is used to trace the bubble edge, and the result is shown in Fig. 4.2d [27].

The bubble area A is represented by the combination of the pixels in the filled area shown in Fig. 4.2c. The circumference is obtained by adding together the pixels constituting the bubble edge. The equivalent bubble diameter d is calculated by:

$$d = 2\sqrt{\frac{A}{\pi}} \tag{4.1}$$

(a) **(b)** **(c)** **(d)**

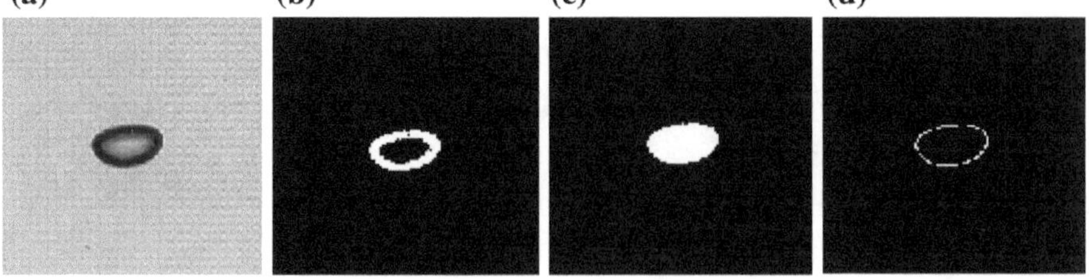

Fig. 4.2 Steps of bubble image processing: **a** raw image, **b** binary image processing and noise suppression, **c** bubble area filling, and **d** bubble edge extraction

The coordinates of the bubble center, x_c and y_c, are given by:

$$x_c = \sum_{i,j \in \Omega} i/N, \quad y_c = \sum_{i,j \in \Omega} j/N \tag{4.2}$$

where i and j are the abscissa and ordinate of the pixel included in the bubble region, respectively. N is the total number of pixels in the area enclosed by bubble edge. Ω is the set of pixels. The bubble trajectory is thereby described with the time-dependent variation of x_c and y_c. Furthermore, the bubble velocity is given by:

$$v_t = \sqrt{\left(v_x^t\right)^2 + \left(v_y^t\right)^2} \tag{4.3}$$

where

$$v_x^t = \left(x_c^{t+\Delta t} - x_c^t\right)/\Delta t \tag{4.4}$$

$$v_y^t = \left(y_c^{t+\Delta t} - y_c^t\right)/\Delta t \tag{4.5}$$

and the superscript t and $t + \Delta t$ represent two consecutive moments.

During the bubble rising process, four kinds of bubble shapes are typical, as shown in Fig. 4.3. Bubble shape depends essentially on two factors, one is viscous effect and the other one is surface tension. As viscous effect is dominant, bubbles are spherical or nearly spherical, and bubble surface is stable and short of

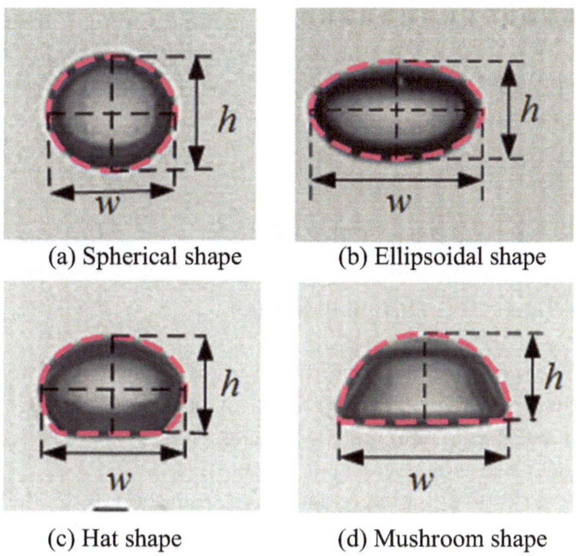

Fig. 4.3 Typical bubble shapes during bubble rising process

oscillation. Otherwise, subjected to overwhelming surface tension effect, bubbles with ellipsoidal, hat or mushroom shapes are common, and bubble wobbling is apparent in this case.

The parameter of bubble aspect ratio, E, is used to describe bubble shape. E is defined as the ratio of bubble height to bubble width,

$$E = h/w \tag{4.6}$$

4.1.1.3 Uncertainty Analysis

The uncertainty in the presented experiment is composed primarily of three constituents, namely the system uncertainty associated with the experiment rig, the uncertainty with bubble size calculation, and the uncertainty entailed by bubble velocity calculation. The components of the experimental rig are fabricated and assembled precisely; therefore the system uncertainty is negligible. Regarding the imaging processing step, according to the assessment in [28, 29], the error with the calculation of equivalent bubble diameter can be quantified with pixels. Here, the attainable spatial resolution of the camera is 0.0879 mm/pixel. The largest bubble diameter is 8.69 mm, which is determined through comparing the bubble diameter obtained through image processing and the physical size of the nozzle. Based on an assumption that the pixel number and bubble size is linearly related, the maximum uncertainty of $\pm 2.78\%$ in bubble size acquisition is deduced. As for bubble velocity calculation, the uncertainty is fostered chiefly in the positioning of the centroid of the same bubble in consecutive bubble images. It is suggested and validated in [30] that the maximum uncertainty in the calculation of the bubble velocity is approximately $\pm 3.19\%$, which is obtained after confirming the maximum deviation of 0.5 pixels in determining bubble centroid through image processing. With a conservative evaluation, the two uncertainties considered should be combined algebraically, so the total measurement uncertainty in the presented experiment is approximately 5.97%.

4.1.2 Experimental Results and Analysis

4.1.2.1 Initial Bubble Shape

The variation of bubble shape immediately downstream of the nozzle is illustrated in Fig. 4.4, where six groups of consecutive images are associated with the six nozzles with different outlet diameters, respectively. The injected air flow rate of 20 mL/h is kept constant for all cases.

Prior to completely detaching from the nozzle, the bubble has a clear bubble tail and a vertically prolonged profile. Then, the bubble oscillates in vertical direction as

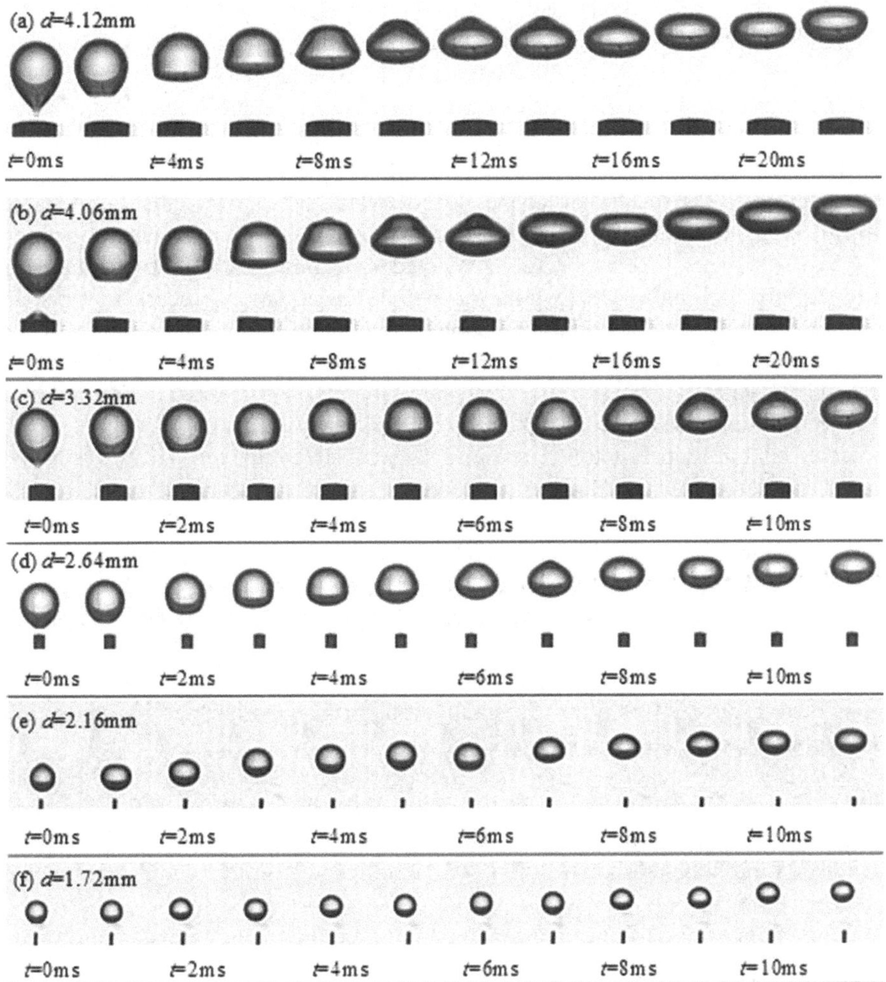

Fig. 4.4 Consecutive bubble images captured with different nozzles at $q_V = 20$ mL/h: **a** $D_0 = 3.00$ mm, $d = 4.12$ mm; **b** $D_0 = 2.20$ mm, $d = 4.06$ mm; **c** $D_0 = 1.60$ mm, $d = 3.32$ mm; **d** $D_0 = 0.60$ mm, $d = 2.64$ mm; **e** $D_0 = 0.19$ mm, $d = 2.16$ mm; **f** $D_0 = 0.16$ mm, $d = 1.72$ mm

the physical contact between the bubble and the nozzle collapses entirely [31]. Here, the equivalent bubble diameter of $d = 2.64$ mm, as shown in Fig. 4.4c, serves as a demarcation between small and large bubbles. For large bubbles, as shown in Fig. 4.4a, b, c, vertical oscillation is apparent and the period lasted is relatively long compared with the three cases of small bubbles. During the experiment, it is observed that the vertical oscillation is apparent even in Region B for large bubbles. The bubble images exhibited in Fig. 4.4 represent typical bubble profiles in the initial stage of bubble evolution after the bubble departs from the nozzle. Since the nozzle exerts a boundary effect on the bubble at the moment it is separated from the nozzle, the lower part of the bubble shrinks and energy is accumulated locally. Eventually, the bubble is separated from the nozzle, but the energy in the lower part of the bubble compels it to oscillate vertically in the ambient water. As the bubble rises continuously, the energy leading to bubble oscillation attenuates. As for large bubbles, such a process is relatively long due to high initial energy from the nozzle. Within this period, bubble shape transforms from initially ellipsoidal shape to hat shape, and then the mushroom shape. After several cycles of oscillation, a stable ellipsoidal bubble shape is accomplished. In contrast, small bubbles shown in Fig. 4.4d, e, f experience a relatively alleviated vertical oscillation and the bubble is only slightly deformed. Meanwhile, the bubble deformation process is short. In particular, in Fig. 4.4f, the spherical bubble is acquired at $t = 2$ ms.

As for large bubbles, with the contribution of surface tension and the influence of ambient water, bubble surface fluctuates violently. In contrast, small bubbles develop into spherical bubbles rapidly after breaking away from the nozzle, which indicates bubble shape evolution attains a balance state relatively easily compared with that of large bubbles. Bubbles produced through ventilation and reducing local static pressure share this characteristic [32].

The variation of the bubble aspect ratio is plotted in Fig. 4.5. For the six cases, the overall variation trends of the bubble aspect ratio are similar. With consistent rising of the bubble, the bubble aspect ratio decreases with fluctuations, signifying the vertical oscillation of bubble surface.

The bubble aspect ratio of large bubble declines sharply within the time span considered. From $t = 0$ to 6 ms, for small bubbles, the spherical and hat shapes are predominant, as is evident in Fig. 4.5. For large bubbles, as shown in Fig. 4.4, they are elongated in vertical direction, then their profiles are flattened apparently. Small bubble secures a small volume and low rising velocity, and the effect of ambient water resistance and surface tension is thereby mitigated. At 6 ms $< t <$ 22 ms, the majority of the bubbles shown in Fig. 4.5 are ellipsoids. In this context, as bubble

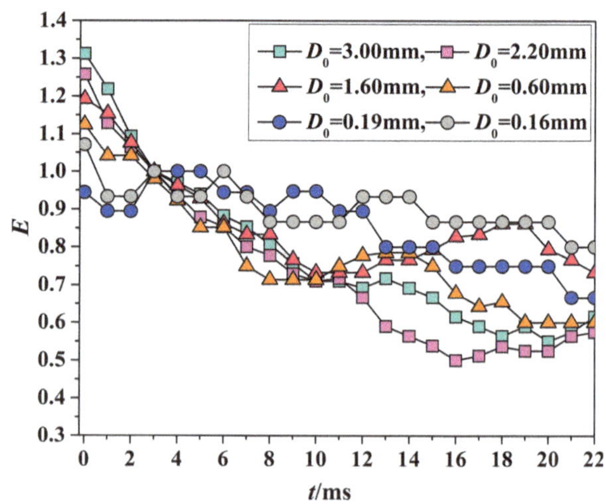

Fig. 4.5 Variation of bubble aspect ratio as bubble rises

velocity increases, the resistance to bubble rising is strengthened. Consequently, bubble velocity is reduced due to the resistance, and the following result is that the resistance dwindles. Therefore, bubble rising is such a process involving the repeated variation of bubble velocity, accompanied by bubble shape oscillation, until the equilibrium state is obtained.

4.1.2.2 Bubble Trajectory

(1) Bubble trajectory in Region A

After the bubble detaches from the nozzle, the buoyant force drives it to ascend in stationary water. Since the DOV plane is two-dimensional, the captured image records the projection of the three-dimensional bubble trajectory onto the DOV plane. According to the experimental results obtained in [30], the bubble diameter values obtained with two-dimensional and three-dimensional methods are nearly equivalent, so the bubble velocity calculated based on two-dimensional bubble images is reliable.

In Region A, bubble rising trajectories with the six nozzles are plotted in Fig. 4.6, where the small circles only indicate transient bubble positions and the identical time interval between neighboring two positions is 5 ms. The flow rate of the injected air is 20 mL/h, as applies to all cases in Fig. 4.6.

It has been proved within this region that the variation of bubble velocity and bubble surface oscillation are remarkable [33]. In Fig. 4.6, near the nozzle outlet, each bubble trajectory includes a linear segment, although the length of the linear segment varies. The largest bubble is associated with the longest linear segment with the length of 79.62 mm, followed by the smallest bubble, as shown in Fig. 4.6a, f. And the length of the latter is about half of that of the former. Then the bubble trajectory begins to wobble and the trajectory shown in Fig. 4.6d is featured by the most violent horizontal deviation with respect to y axis. It is noticeable that the largest bubble rises along a nearly straight route. On the whole, the trajectory oscillation of large bubbles is relatively weak and this trend is in accordance with the results obtained in [34].

Within Region A, bubble velocity components, V_x, and V_y are obtained through image processing, and the results are plotted in Fig. 4.7. The variation of bubble aspect ratio is plotted in Fig. 4.7 as well. At this stage, the bubble is in a state of instability, as can be inferred from the fluctuation of velocity curves. At bubble equivalent diameters of 1.72, 2.16 and 2.64 mm, vertical velocity component increases rapidly near the nozzle. While as for other three bubble equivalent diameters, bubble motion exhibits strong randomness, vertical velocity varies drastically as the bubble progresses and gets stable after several oscillation periods.

Of interest is the correlation between the bubble aspect ratio and bubble velocity. The two parameters vary synchronously but in an inverse manner. The crest of the bubble aspect ratio curve corresponds to the trough of the vertical velocity curve, as shown in Fig. 4.7. Such a state is also proved in [17]. As vertical bubble velocity increases, the

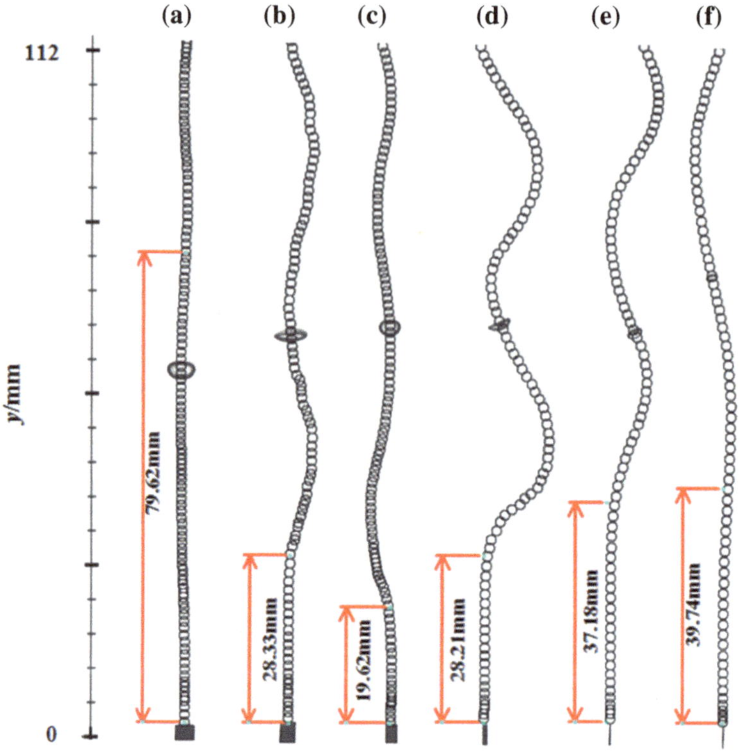

Fig. 4.6 Bubbles trajectories with different nozzles: **a** $D_0 = 3.00$ mm, $d = 4.12$ mm; **b** $D_0 = 2.20$ mm, $d = 4.06$ mm; **c** $D_0 = 1.60$ mm, $d = 3.32$ mm; **d** $D_0 = 0.60$ mm, $d = 2.64$ mm; **e** $D_0 = 0.19$ mm, $d = 2.16$ mm; **f** $D_0 = 0.16$ mm, $d = 1.72$ mm

resistance from ambient water increases as well. Thus there is a large pressure difference between the upper and lower sides of the bubble, so the bubble tends to be flattened and the bubble aspect ratio decreases accordingly. Since the flattened bubble suffers from a large drag force, the increase of vertical bubble velocity is hindered. These alternate steps boost the bubble oscillation as it rises. As for large bubbles, there is no consistent relationship between the bubble aspect ratio and velocity components, as indicated in Fig. 4.7. The variation of the bubble aspect ratio with bubble rising is severe in Fig. 4.7a, b relative to that in Fig. 4.7c, f. The aspect ratio in Fig. 4.7c is linked with a nearly spherical bubble. Regarding the bubble variation of horizontal bubble velocity with bubble rising, Figs. 4.7c and 4.6c are in good accordance, and small horizontal velocity magnitude is compatible with the linear bubble trajectory.

(2) Bubble trajectory in Region B

Bubble trajectories in Region B are shown in Fig. 4.8. The bubble trajectories in Region B are characterized by periodicity, and all the six trajectories can be

Fig. 4.7 Variation of bubble aspect ratio and velocity components as bubble rises in Region A: **a** $D_0 = 3.00$ mm, $d = 4.12$ mm; **b** $D_0 = 2.20$ mm, $d = 4.06$ mm; **c** $D_0 = 1.60$ mm, $d = 3.32$ mm; **d** $D_0 = 0.60$ mm, $d = 2.64$ mm; **e** $D_0 = 0.19$ mm, $d = 2.16$ mm; **f** $D_0 = 0.16$ mm, $d = 1.72$ mm

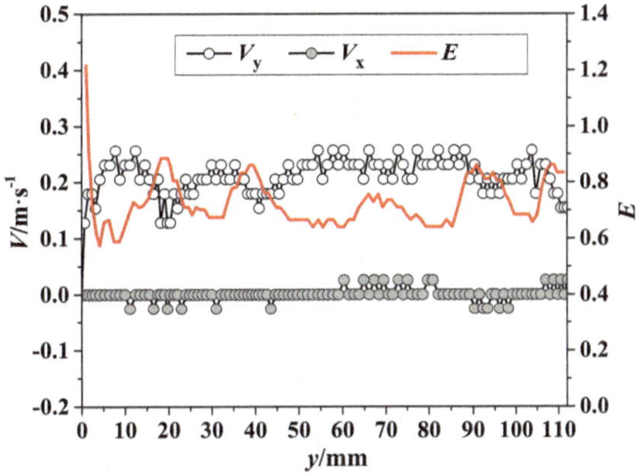

(a) D_0=3.00mm, d=4.12mm

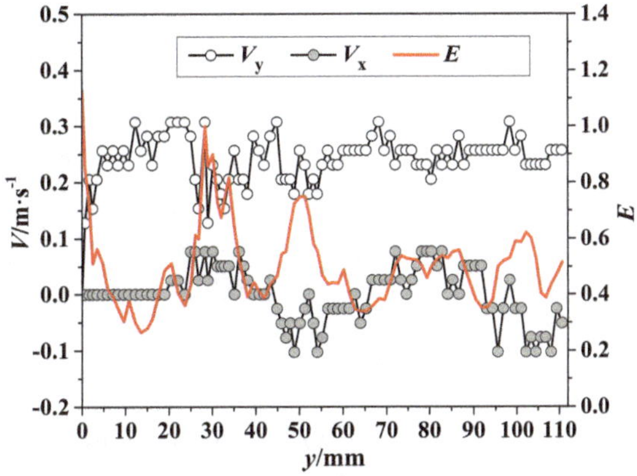

(b) D_0=2.20mm, d=4.06mm

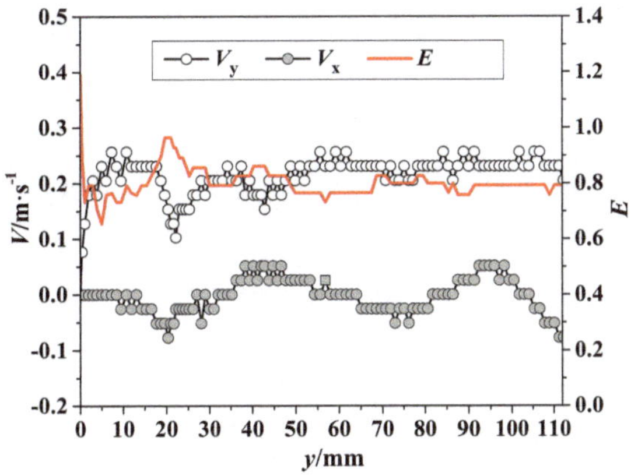

(c) D_0=1.60mm, d =3.32mm

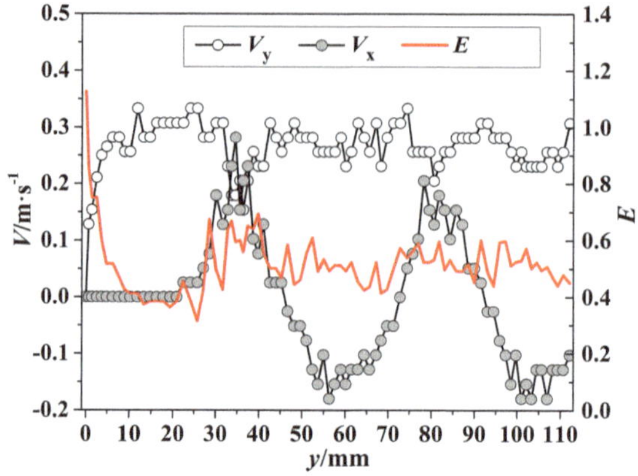

(d) D_0=0.60mm, d =2.64mm

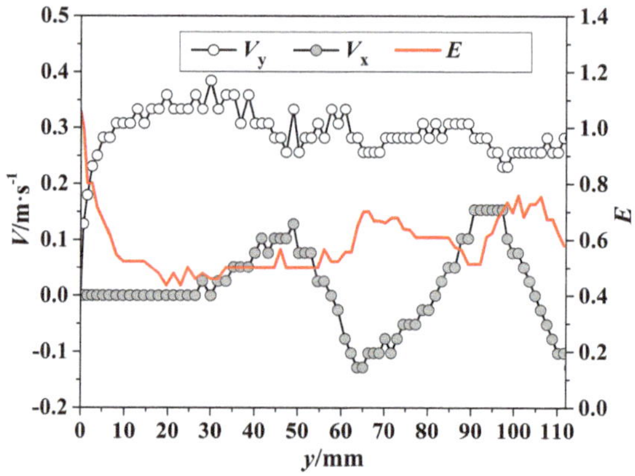

(e) D_0=0.19mm, d =2.16mm

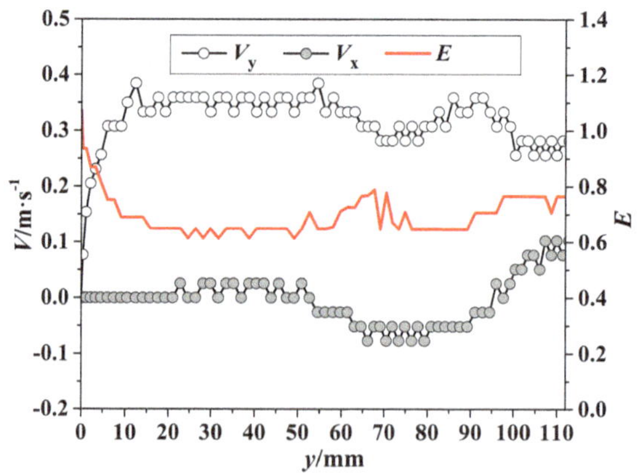

(f) D_0=0.16mm, d=1.72mm

Fig. 4.7 (continued)

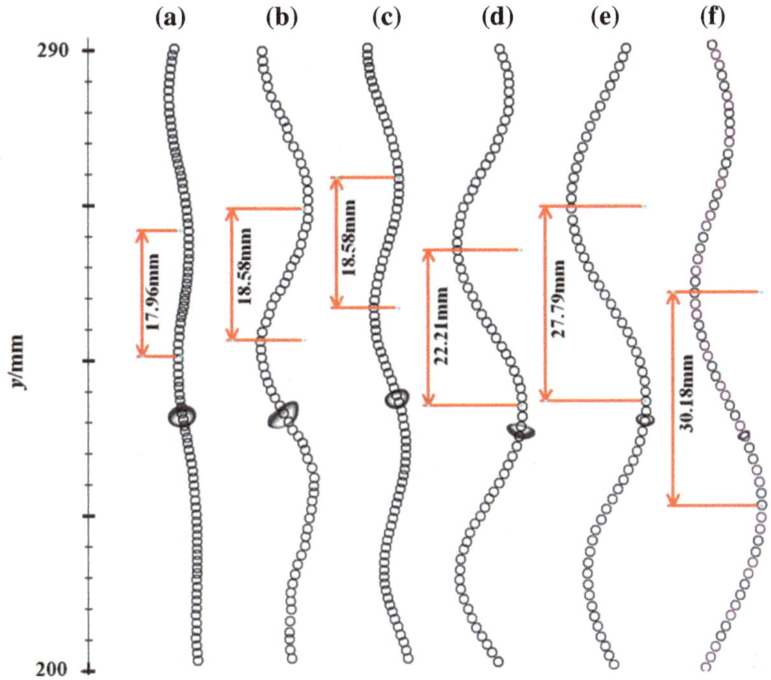

Fig. 4.8 Bubbles trajectories with different nozzles: **a** $D_0 = 3.00$ mm, $d = 4.12$ mm; **b** $D_0 = 2.20$ mm, $d = 4.06$ mm; **c** $D_0 = 1.60$ mm, $d = 3.32$ mm; **d** $D_0 = 0.60$ mm, $d = 2.64$ mm; **e** $D_0 = 0.19$ mm, $d = 2.16$ mm; **f** $D_0 = 0.16$ mm, $d = 1.72$ mm

approximated with the sinusoidal function. The largest bubble is associated with a fairly straight trajectory, and the largest sinusoidal wavelength is attached to the smallest bubble. As the equivalent bubble diameter decreases, the sinusoidal wavelength decreases consistently.

The variation of the bubble aspect ratio and bubble velocity in Region B is plotted in Fig. 4.9. It is obvious that the bubble aspect ratio still varies, particularly for large bubbles. Furthermore, the periodicity of bubble aspect ratio variation is explicit, as differs significantly from the situations indicated in Fig. 4.7. In contrast, in Fig. 4.9e, f, bubble deformation is relatively slight. Apparently, the fluctuation of vertical bubble velocity is suppressed. Moreover, it is seen that the horizontal velocity curves are similar to bubble rising trajectories displayed in Fig. 5.8, such a correspondence is also proved in [35]. Bubbles situated in Region B are exposed to a stable state under which the bubble motion is dominated by horizontal bubble velocity. Moreover, in Region B, the disturbance of vertical bubble velocity fluctuation to bubble trajectory is weakened evidently.

Based on the above analysis, the bubble rising process covers three distinct stages: (1) Vertical bubble velocity dominated stage, at which vertical bubble velocity varies drastically while horizontal bubble velocity is suppressed, and

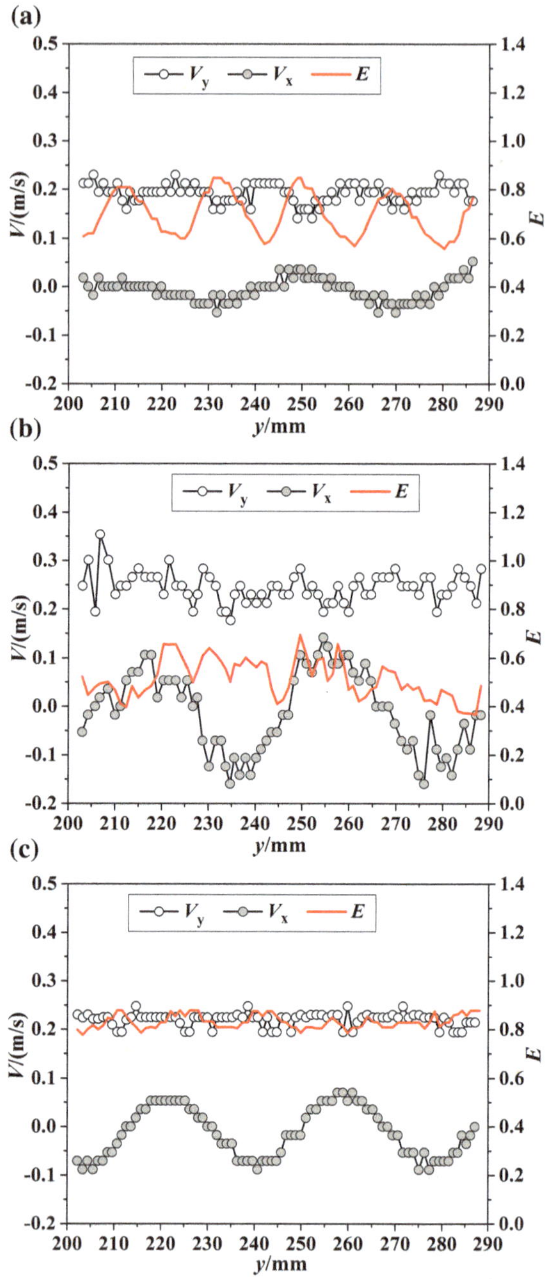

Fig. 4.9 Variation of bubble aspect ratio and velocity components as bubble rises in Region B: **a** $D_0 = 3.00$ mm, $d = 4.12$ mm; **b** $D_0 = 2.20$ mm, $d = 4.06$ mm; **c** $D_0 = 1.60$ mm, $d = 3.32$ mm; **d** $D_0 = 0.60$ mm, $d = 2.64$ mm; **e** $D_0 = 0.19$ mm, $d = 2.16$ mm; **f** $D_0 = 0.16$ mm, $d = 1.72$ mm

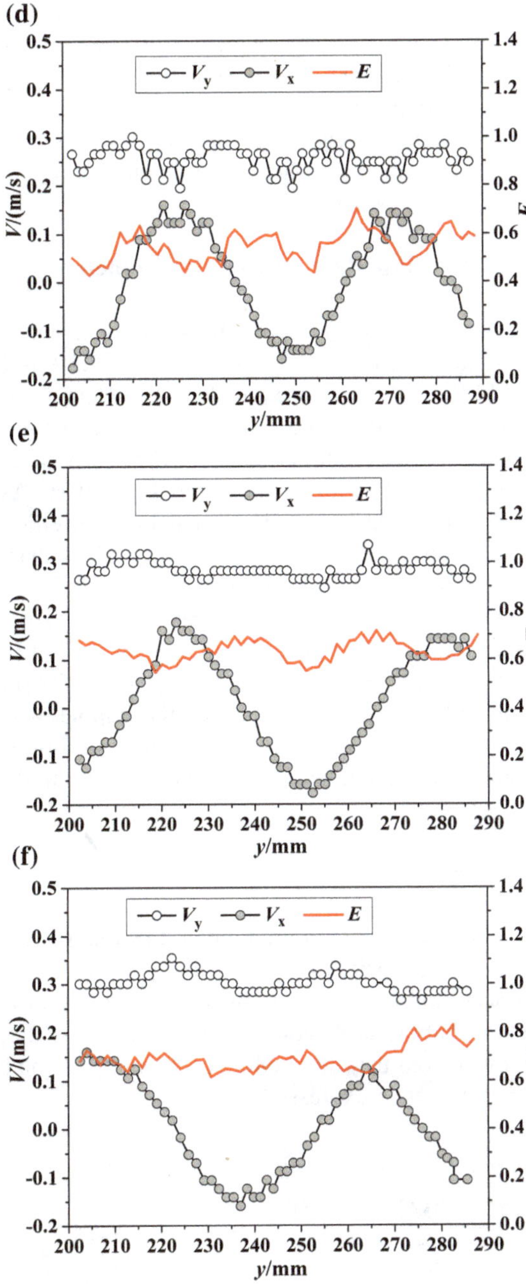

Fig. 4.9 (continued)

bubble rises along a fairly straight route; (2) Bubble velocity fluctuating stage, at which the bubble begins to rise in a zigzag pattern, fluctuations of both horizontal and vertical bubble velocities are prevalent; (3) Stable bubble rising stage, at which bubble trajectory is approximated with the sinusoidal function and horizontal bubble velocity is predominant. Regarding bubbles with different initial size, the time spans for the same stage are significantly different.

4.1.2.3 Influence of Air Flow Rate on Bubble Geometry

With the same nozzle, the adjustment of the flow rate of injected air leads to the variation in bubble initial size. Here, the nozzle of $D_0 = 2.20$ mm is selected as a representative and bubbles with various equivalent diameters produced thereby are exhibited in Fig. 4.10. From Fig. 4.10a–f, q_V increases consistently. The bubble evolution is at the vertical velocity dominated stage. The nearly spherical bubble with the equivalent diameter of 1.13 mm is shown in Fig. 4.10a. As the equivalent diameter increases, the spherical bubble is shifted into the ellipsoidal one, as is demonstrated in Fig. 4.10b, c. From Fig. 4.10c, d, bubble shape experiences a remarkable change and the ellipsoidal bubble is replaced with a flattened one. Meantime, the oscillation of bubble is reinforced. As q_V increases further, the equivalent bubble diameter increases to 8.48 mm, as displayed in Fig. 4.10f. Such a large-size bubble can be produced only in stationary water, for flowing water, the shear effect will outweigh the surface tension of the bubble, and bubble surface integrity cannot sustain. A global view of Fig. 4.10 indicates bubble shape irregularity as well as the oscillation of the bubble surface is enhanced as the air flow rate increases.

4.1.2.4 Terminal Velocity

As the drag and buoyancy forces exerted upon the rising bubble are entirely counteracted in y direction, vertical bubble velocity no longer changes and the terminal velocity is accomplished. Such a distinct situation represents an equilibrium state for the rising bubble in stationary water, signifying that the bubble has been completely relieved from the initial effect of the nozzle. The balance between the drag and buoyancy forces is expressed as:

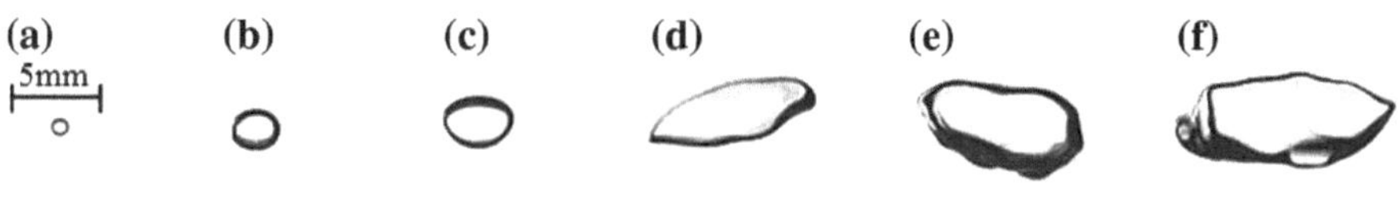

Fig. 4.10 Comparison of bubble geometry at different air flow rate with the nozzle of $D_0 = 2.20$ mm: **a** $d = 1.13$ mm, $q_V = 2$ mL/h; **b** $d = 2.72$ mm, $q_V = 8$ mL/h; **c** $d = 3.89$ mm, $q_V = 16$ mL/h; **d** $d = 6.01$ mm, $q_V = 100$ mL/h; **e** $d = 7.43$ mm, $q_V = 400$ mL/h; **f** $d = 8.48$ mm, $q_V = 800$ mL/h

$$C_D \frac{1}{2} \rho_l V_T^2 \frac{\pi d^2}{4} = (\rho_l - \rho_g) g \frac{\pi d^3}{6} \tag{4.7}$$

where C_D is the drag coefficient, the terminal velocity V_T is thus given by:

$$V_T = \sqrt{\frac{4(\rho_l - \rho_g) g d}{3 C_D \rho_l}} \tag{4.8}$$

Tomiyama et al. predict that the value of C_D for the distorted bubbles is larger than that of spherical bubbles and therefore recommend the following expressions for C_D in pure liquid [36]:

$$C_D = \max\left\{ \min\left[\frac{16}{Re}\left(1 + 0.15\, Re^{0.687}\right), \frac{48}{Re} \right], \frac{8}{3} \frac{Eo}{Eo + 4} \right\} \tag{4.9}$$

where Eo is the Eötvös number and is defined as:

$$Eo = \frac{g(\rho_l - \rho_g) d^2}{\sigma} \tag{4.10}$$

The Weber number We is given by:

$$We = \frac{\rho_l V_T^2 d}{\sigma} \tag{4.11}$$

In addition to calculating bubble terminal velocity through C_D, the approach of directly predicting such a velocity has been proposed by Mendelson et al. [37]. They associate bubble rising velocity with hydrodynamic principles of waves and construct the following equation:

$$V_T = \sqrt{\frac{2\sigma}{d(\rho_l + \rho_g)} + \frac{(\rho_l - \rho_g)}{\rho_l} \frac{gd}{2}} \tag{4.12}$$

As a prerequisite of statistical investigation of bubble geometry and bubble velocity, a large group of bubble data is obtained in Region B through adjusting the flow rate of the injected air. The bubble terminal velocity and corresponding equivalent bubble diameter are acquired and processed statistically. The result is shown in Fig. 4.11. At $d < 1.2$ mm, a nearly linear relationship is proved between bubble terminal velocity and equivalent bubble diameter. The variation tendency of bubble terminal velocity distribution at this stage agrees well with the result obtained through Eqs. (4.8) and (4.9), but is contrary to Mendelson's prediction with Eq. (4.12). In this context, small bubbles are nearly spherical and fluctuate only slightly so that the motion of small bubbles does not conform to the hydrodynamic wave theory. Therefore Eq. (4.12) is only suitable for conditions of

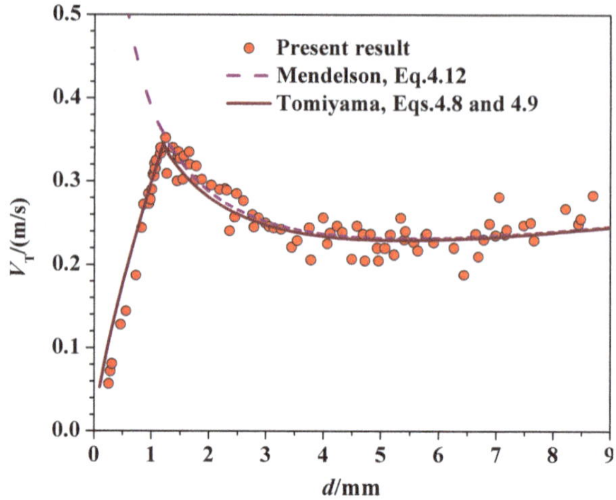

Fig. 4.11 Variation of bubble terminal velocity with equivalent bubble diameter in Region B

medium-sized bubbles. At 1.2 mm $\leq d < 3.5$ mm, with the increase of d, V_T decreases gradually. At 3.5 mm $\leq d < 8.6$ mm, data points of V_T are well situated near the two empirical curves.

As bubble size increases, the deviation between bubble terminal velocity magnitudes is intensified, as attributes to bubble surface deformation. The statistical result in Fig. 4.11 demonstrates that the effect of nozzle outlet diameter is negligible as the rising bubble reaches the state of equilibrium. Furthermore, although the bubble geometry is complex, in particular for large bubbles, the variation of bubble terminal velocity with equivalent bubble diameter is statistically stable.

Tomiyama et al. argue that the bubble aspect ratio is closely related to bubble terminal velocity and then propose the following relationship [38]:

$$V_T = \frac{\sin^{-1}\sqrt{1-E^2} - E\sqrt{1-E^2}}{1-E^2}\sqrt{\frac{8\sigma}{\rho_l d}E^{4/3} + \frac{(\rho_l - \rho_g)}{2\rho_l}\frac{E^{2/3}}{1-E^2}} \tag{4.13}$$

Here, the bubble terminal velocity obtained in the present study and the result predicted through Eq. (4.13) are comparatively shown in Fig. 4.12. Two equivalent bubble diameters, 4.12 and 1.72 mm are selected representatively. Regarding the large bubble, as shown in Fig. 4.12a, the two velocity curves are in good agreement. The present experimental result supports the feasibility of Eq. (4.13) in predicting terminal velocity. In contrast, as for the two velocity curves associated with the small bubble in Fig. 4.12b, the imparity is clear. In this context, according to the analysis made by Tomiyama, the error of Eq. (4.13) can be reduced by increasing the accuracy of bubble aspect ratio calculation or by considering additional effect of the external forces acting on the bubble, such as the virtual mass force and lift force.

Fig. 4.12 Variation of bubble terminal velocity and aspect ratio with bubble size

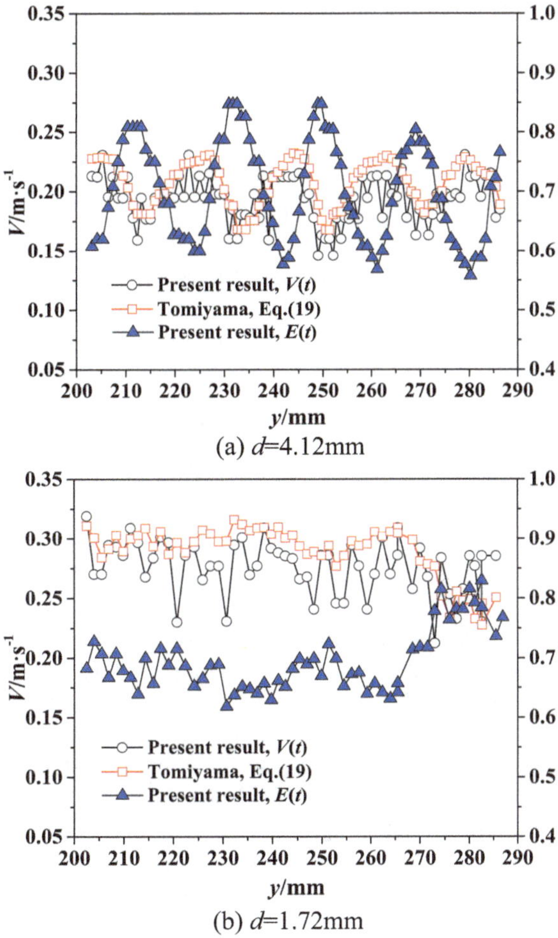

Variations of the bubble aspect ratio and velocity are synchronous in both Fig. 4.12a, b. Such a situation is similar to that illustrated in Fig. 4.9. Nevertheless, in Fig. 4.9, the velocity component V_y, instead of the resultant velocity, varies synchronously with the bubble aspect ratio. Therefore, in the state of equilibrium, the vertical bubble velocity is dominant irrespective of the effect of bubble size. In consideration of the consistency of bubble velocity shown in Fig. 4.12, Eq. (4.13) is reliable in the prediction of bubble velocity provided that accurate data of the bubble aspect ratio are available.

4.1.2.5 Bubble Aspect Ratio

The bubble aspect ratio is often linked with dimensionless parameters such as Reynolds number *Re*, Weber number *We* and Morton number *Mo* [39].

The relationship between Tadaki number and the bubble aspect ratio suffers from uncertainties in the presence of small viscous force [40]. Moore et al. derive a relationship between *We* and *E* based on the balance between the dynamic pressure and the capillary pressure [41]:

$$We = \frac{4E^{4/3}(E^{-3} + E^{-1} - 2)}{(E^{-2} - 1)^3}\left[E^{-2}\sec^{-1}E^{-1} - \left(E^{-2} - 1\right)^{1/2}\right]^2 \tag{4.14}$$

While Taylor and Acrivos propose another model based on Eq. (4.14) [42]:

$$E = \frac{1}{1 + 5We/32} \tag{4.15}$$

Sugihara et al. propose a correlation between *E* and *We* based on Moore solution to make it suitable at large Weber number [43]:

$$E = \left[\frac{1}{1 + 9We/64} + \frac{0.04We^2}{(3.7 - We)^{0.5}}\right]^{-1} \tag{4.16}$$

The relationship between *E* and *We* is described in Fig. 4.13. As $We < 0.1$, the data of *E* are rather stable and approach 1, and bubbles are nearly spherical within this range of Weber number. As $0.1 < We < 1.03$, *E* is slightly less than 1, and spherical shape is slightly impaired. With respect to these two Weber number ranges, the statistical results obtained in this study agree well with the results of Eqs. (4.14), (4.15) and (4.16). The experimental techniques and image-processing algorithms allow the acquisition and depiction of bubbles at small Weber number, and the statistical results offer a sound support to the validation of the three equations. At $We > 1.5$, the data points of *E* obtained in the present study are

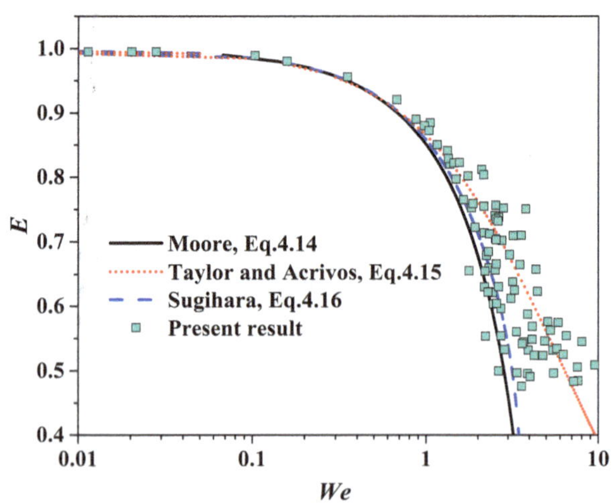

Fig. 4.13 Statistical relationship between bubble aspect ratio and Weber number

scattered. E drops sharply within this range as We increases from an overall viewpoint. Particularly, near the Weber number of 3.2, the bubble aspect ratio is sensitive to Weber number, a slight variation of Weber number will trigger a remarkable increase or decrease in the bubble aspect ratio. Approaching the Weber number of 10, the feasibility of Eq. (4.15) is evident relative to the other two equations.

The relationship between Tadaki number Ta and Re and Mo is given by:

$$Ta = ReMo^{0.23} \tag{4.17}$$

Meanwhile, Tadaki et al. propose a relationship between Ta and E based upon the analysis of a large amount of experimental data [44]:

$$E^{1/3} = \begin{cases} 1 & Ta < 2 \\ 1.14Ta^{-0.176} & 2 < Ta < 6 \\ 1.36Ta^{-0.28} & 6 < Ta < 16.5 \\ 0.62 & 16.5 < Ta \end{cases} \tag{4.18}$$

Myint et al. measure the bubble aspect ratio in infinite stagnant liquids and propose another model for correlating Ta and E [45]:

$$E = 1 - 0.0487Ta - 0.0289Ta^2 \tag{4.19}$$

The relationship between E and Ta is described in Fig. 4.14. As $Ta < 0.8$, the values of E remain rather stable and approach 1, and bubbles are nearly spherical within this range of Tadaki number. The results of Eq. (4.18) agree well with the data obtained in the present study. As $0.8 < Ta < 2.1$, with the increase of Ta, E decreases sharply, and ellipsoidal bubbles are predominant in this case. The results of Eq. (4.19) are close to the data obtained here. As $Ta > 2.1$, the deviation

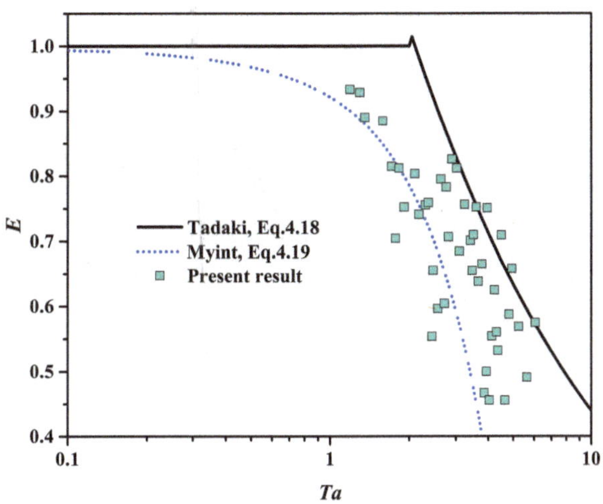

Fig. 4.14 Statistical relationship between bubble aspect ratio and Tadaki number

between the data points of E is promoted, and bubble deformation is remarkable. Near the Tadaki number of 3.8, the bubble aspect ratio is sensitive to Tadaki number, as is similar to the situation faced by Weber number.

Based on the analysis of Weber number and Tadaki number, it is evident that the relationship between the bubble aspect ratio and Weber number or Tadaki number cannot be represented with a monotonously declining curve. The swarming of data points implies that the surface tension, bubble geometry and bubble terminal velocity are fluctuating interrelatedly over some balance state. The fluctuations are the most essential reflection of bubble dynamics. In both Figs. 4.13 and 4.14, the scattered data points still cannot be covered with a simple variation curve. This manifests that the bubble evolution is a multi-factor influenced process, both bubble morphology and the response of bubble to external influence are featured by high diversity.

The possibility of constructing a generalized graphical correlation involving *Eo*, *Mo* and *Re* is sought in [46]. Later, Tomiyama et al. propose a prediction model to adapt to a wide range of *Eo* and *Mo* [36]. The Grace diagram based on the results of the present study is plotted in Fig. 4.15, where a wide range of *Eo*, i.e., $0.88 \times 10^{-2} < Eo < 1.02 \times 10^{1}$ and $Mo = 2.168 \times 10^{-10}$ are selected. The data points obtained in the present study are mainly distributed in the regimes of spherical bubble, ellipsoidal bubble and wobbling bubble. It is clear that bubbles are spherical at relatively low values of *Re* and *Eo*, the wobbling regime is matched with large *Re* and intermediate *Eo* values. Meanwhile, the ellipsoidal regime corresponds to intermediate values of *Re* and *Eo*, and both *Eo* and *Re* are high in the spherical-cap regime. Moreover, the result calculated with the model suggested by Tomiyama is in accordance with the result obtained in the present study, indicating that statistically stables results of the equivalent bubble diameter, the bubble aspect ratio and terminal velocity of the bubble have been achieved.

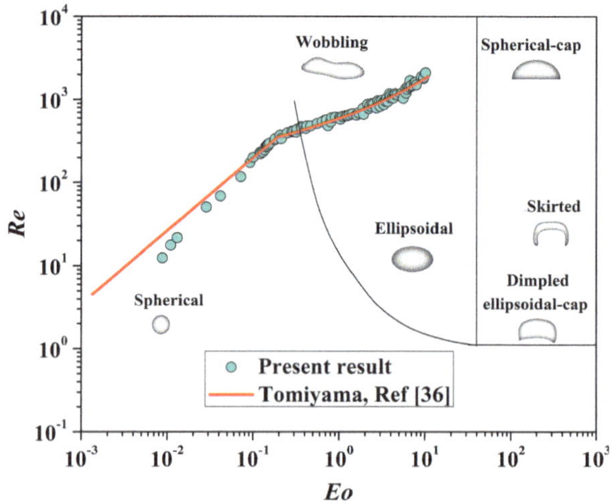

Fig. 4.15 Correspondence between bubble shape and dimensionless numbers

4.1.3 Summary

(1) The evolution of the rising bubble in stationary water depends significantly on bubble initial size. Large bubble is featured by the transition from ellipsoidal shape, hat shape, mushroom shape and finally to ellipsoidal shape. Small bubble attains spherical shape rapidly after it detaches from the nozzle. With the same nozzle, the change in the air flow rate enables the production of diverse bubble size and bubble geometry. Bubble surface oscillation is reinforced as bubble size increases.
(2) The bubble rising process involves three consecutive stages, vertical bubble velocity dominated stage, bubble velocity fluctuating stage and stable bubble rising stage. Characteristics of bubble velocity fluctuations and bubble trajectory geometry are responsible for the demarcation. Large bubble is associated with relatively mitigated horizontal bubble velocity variation and smooth bubble trajectory. All bubbles share the sinusoidal type trajectory at the stable rising stage.
(3) Statistical relationship between bubble terminal velocity and bubble size is depicted based on experimental results. As bubble size increases, bubble terminal velocity undergoes a turning point before attaining a fairly stable state. It is evident that surface tension, bubble geometry and bubble terminal velocity vary interrelatedly. Bubble shape diversity is reflected by the Grace diagram, which demonstrates the correspondence between bubble shape and Reynolds and Eötvös numbers.

4.2 Bubbles Released in Horizontal Water Flow

The diversity of bubbly flows is related to the interaction between bubbles and the surrounding liquid. Bubbles can be generated via injecting air into liquid, which has been witnessed in Sect. 4.1. Such a strategy has been implemented in hydraulic and environmental engineering to improve the content of the dissolved oxygen in water, therefore providing a favorable environment for fishes and other aquatic animals [47]. In this context, bubble size and the distribution of the bubbles play an important role in the enhancement of the mass transfer between the bubble and the liquid phase.

The liquid flow imposes significant effects on the trapped bubbles. In shear flows, the bubble would deform due to the velocity gradients of the liquid flow [48]. Regarding the rotating flow, it influences the bubble through the centrifugal force, pushing the bubble move away from the center of rotation [49]. Provided that bubbles travel in a thin pipe or duct, the wall effect would inevitably influence the bubble velocity distribution as well as the bubble geometry [50]. In particular, the response of the bubbles near the wall to the viscous sub-layer results in distinct features in terms of the bubble size and bubble shape relative to bubbles located elsewhere. Taking the breakup of bubbles into account, the treatment of the bubbly flow would be rather difficult [51]. As has been reported in publicized literature, both measurement techniques and numerical simulation have been used in the bubbly flow investigation. In comparison, measurement results such as bubble size distribution are more practical [52]. Numerical models for depicting the motion of bubbles have been established but the influential factors cannot be perfectly considered [53]. Furthermore, as the bubble shape deviates considerably from the spherical form, the forces exerted on the bubble surface might not be well predicted. Subsequently, the bubble motion obtained numerically differs from the fact [54].

Regarding previous studies on the bubbles generated via ventilation, the emphasis is usually placed on bubble plumes instead of individual bubbles [55]. Hitherto, the observation of individual bubbles has been performed overwhelmingly with the stagnant liquid. Trapped in the liquid flow, a bubble, along with bubble velocity components, is schematically shown in Fig. 4.16. As can be seen, the motion of the bubble is determined by multiple effects, which result from the forces exerted on the bubble. Among those forces, the shear force due to the velocity difference between the bubble and the liquid is dominant [56]. The bubble velocity component, u_b, depends on such a shear force. This force increases with liquid velocity. The slip velocity, v_b, could be obtained through the devised formula [57]. The velocity, $\vec{v}_w$, arises due to the wake effects of the bubble; but it was often neglected for small bubble size. It is appreciable in Fig. 4.16 that the direction and magnitude of the resultant bubble velocity, $\vec{v}$, are influenced by not just liquid flow characteristics but also the properties of the bubble itself.

In this section, the purpose is to obtain bubble characteristics as the bubble travels in the liquid cross flow. A water tunnel is used to provide uniform water flow. Individual bubbles are produced in the water flow using the ventilation technique. High-speed photography is used to capture instantaneous bubble images. An image-processing code is developed to distinguish the bubble and to collect bubble data from the bubble images. The effects of the water flow velocity and the flow rate of the injected air on the bubble are jointly considered. The variations in

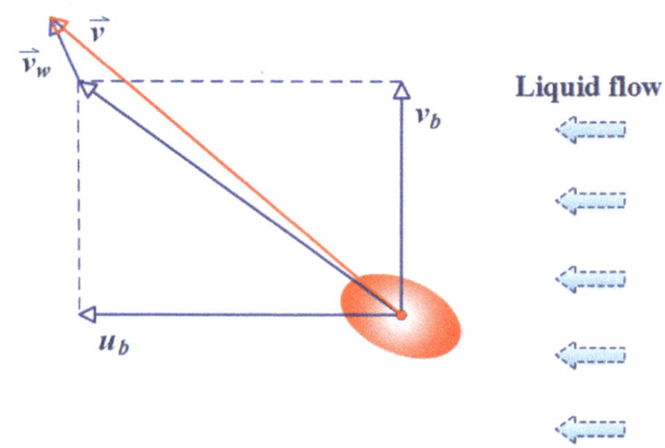

Fig. 4.16 Velocity components of the bubble in liquid flow

bubble geometry and the bubble velocity with the motion of the bubble are to be explained. Moreover, the surface tension and the buoyant force associated with the bubble are anticipated to be analyzed through non-dimensional numbers; hence, an extension with respect to the existing empirical relationship, which is constructed solely for the bubble in stagnant water, would be attained.

4.2.1 Experimental Preparations

4.2.1.1 Water Tunnel and Ventilation Apparatus

A water tunnel, schematically shown in Fig. 4.17, is used to provide steady and uniform liquid flows [58]. The observation of the bubble will be performed in the test section of the water tunnel. The dimensions of the test section are 700 mm × 50 mm × 315 mm. All the four side walls of the test section are made of acrylic plexiglass. The water velocity magnitude in the test section is controlled through adjusting the frequency of the motor that drove the pump.

Pure water of 22 °C is circulated in the loop shown in Fig. 4.17. Prior to the experiment, the cross-sectional velocity distributions in the test section are measured

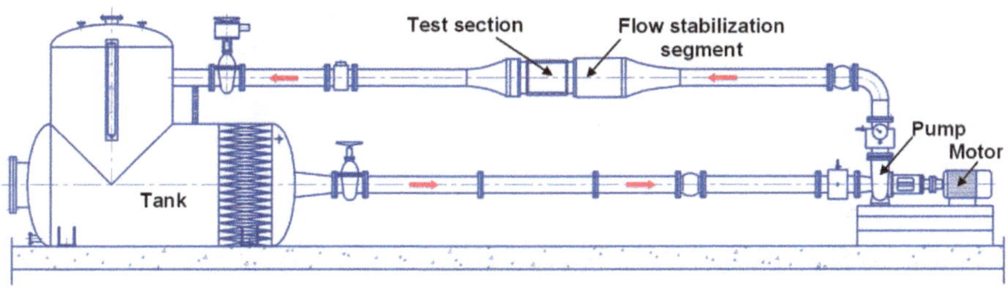

Fig. 4.17 Schematic diagram of the water tunnel

Fig. 4.18 Configuration of PIV components

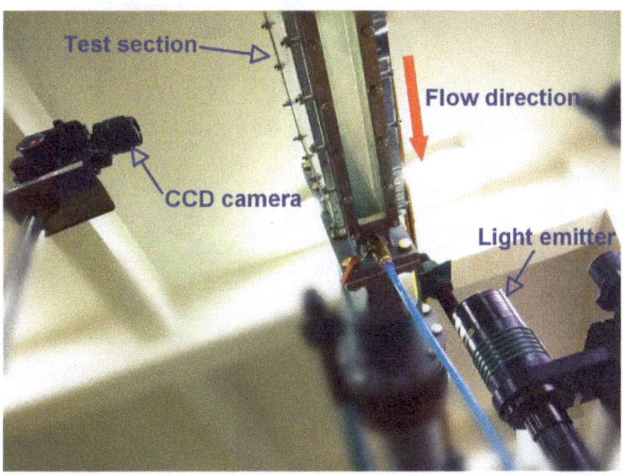

using the PIV technique, and the PIV system manufactured by LaVision company is used. Major components of the system are exhibited in Fig. 4.18. An Nd:YAG laser with the light wavelength of 532 nm is used to illuminate the water flow. A CCD camera of Imager Pro SX 5M with an image resolution of 2456 × 2058 pixels is used. The maximum image-capturing frequency is set to 14.5 fps. Hollow glass particles with diameters ranging from 20 to 50 μm are used as tracing particles. Data acquisition and processing are conducted using the Davis software.

Based on the averaged velocity measurement results, the streamline patterns over the cross section parallel to the bulk flow direction are obtained and displayed in Fig. 4.19. The flow direction in Fig. 4.19 is from right to left; and the coordinates

Fig. 4.19 Streamline patterns in the test section

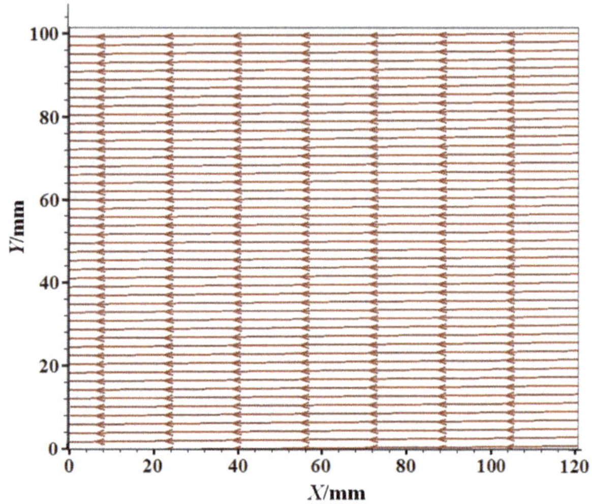

(a) f=7.05 Hz

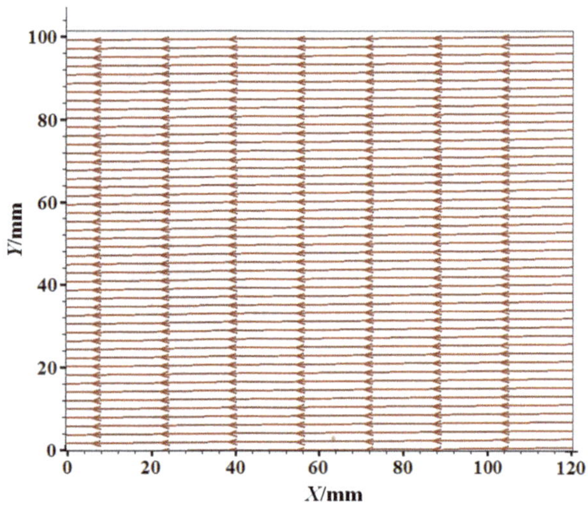

(b) f=7.95 Hz

of $Y = 0$ mm overlaps with the horizontal symmetry centerline of the test section. As can be seen, for the two motor frequencies, $f = 7.05$ and 7.95 Hz, water velocity distributions are uniform in both streamwise and transverse directions in the test section, furnishing a favorable environment for bubble experiments.

Further PIV experiments are conducted to examine the operation stability of the water tunnel. For different motor frequencies between $f = 7.05$ and 7.95 Hz, water velocity in the test section is measured. The variation in the water velocity magnitude, u_l, as a function of f is plotted in Fig. 4.20. As can be seen, the correspondence between u_l and f is well organized. Moreover, the data points displayed in Fig. 4.20 can be well fitted with a linear relationship, which is formulated in Fig. 4.20 as well. Therefore, stable operation of the water tunnel is proven; furthermore, the comparison of bubble properties for different water flow velocities is thereby possible. Turbulent fluctuations in the test section are suppressed via the stabilization segment deployed upstream of the test section. In this context, the influence of turbulent fluctuations on the bubble is beyond the scope of this section [59].

A circular brass tube, with an inner diameter of 3 mm and out diameter of 5 mm, serves as a nozzle and is installed in the test section for discharging air into the water tunnel. The axis of the nozzle is vertical to the water flow direction, and the nozzle outlet is positioned at $Y = 0$ mm. A syringe pump is used to drive continuous injection of air into the water flow. The flow rate of the injected air, q_{sV}, is adjustable. The configuration of the ventilation components is shown in Fig. 4.21. It should be noted that the air-injection manner has a remarkable effect on bubble geometry and the bubble distribution density [60]. The air-injection approach adopted here ensures that bubbles are released in the same plane. Meanwhile, with the cooperation between the flow rate of the injected air and the water flow velocity, the overlapping between bubbles is avoided. In comparison, bubble plumes have been reported in previous ventilation studies [61]. In that case, the bubble-bubble

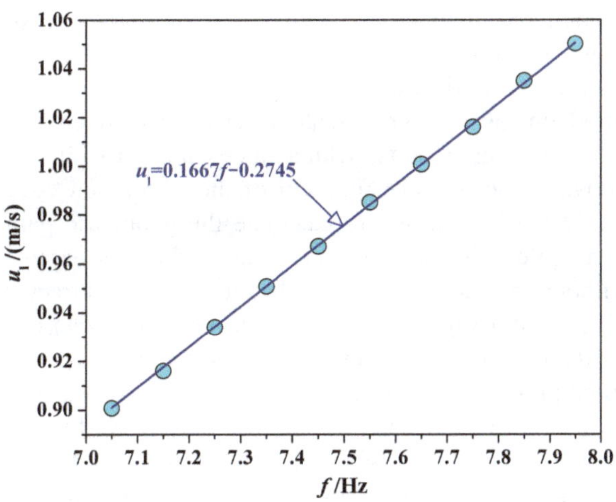

Fig. 4.20 Variation of water velocity magnitude with the motor frequency

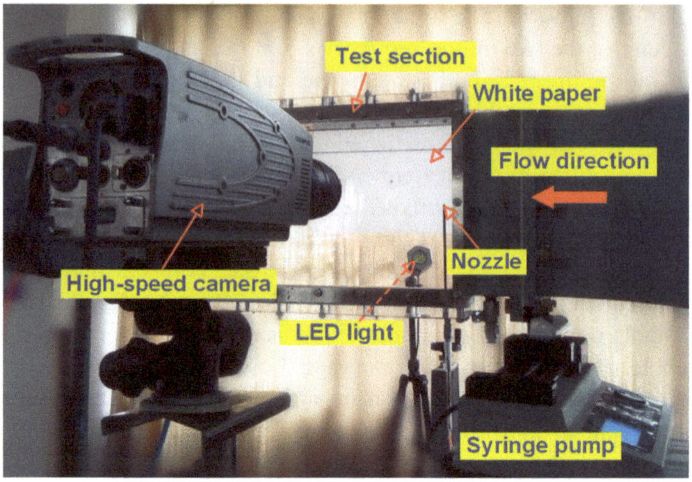

Fig. 4.21 In-situ image of high-speed photography experiment

interaction cannot be neglected; and bubbly flow patterns can be recognized merely from an overall view.

Bubble images were recorded using an OLYMPUS I-SPEED 3 high-speed camera in conjunction with an OLYMPUS ILP-2 light source, as shown in Fig. 4.21. The DOV plane is set to overlap with the plane passing through the axis of the nozzle. The light passes through a white paper before penetrating the water flow. A white plate is used to uniformize the incident light onto the DOV plane; thereby, the monitored window is dominated by uniform distribution of light intensity. Considering the magnitude of the bubble velocity, the exposure time and the frame rate of the camera are set as 0.5 ms and 3000 fps, respectively.

To extract quantitative bubble information from the bubble images captured, an image-processing code is developed based on the commercial MATLAB software. The major steps incorporated in the code are displayed in Fig. 4.22.

In the captured images, the gray scale values associated with the bubble are different from the remaining parts, which opens the possibility of isolating the bubble in the image. The primary function of the developed code in the present study is to identify the bubble and to calculate the geometric parameters and the velocity of the bubble. The background image, which is captured at the same position but without the participation of the bubble, is subtracted from the raw bubble image. The contrast in the image is improved though an image binarisation algorithm. Then the median filter algorithm is used to smooth the image. As one of the most important procedures, the edge of the bubble is traced and extracted using the Canny algorithm [62]. A linear smoothing filter is used for reducing noise for the extracted bubble edge. After the processing of consecutively captured bubble images for the same monitored window, instantaneous bubble edges are obtained

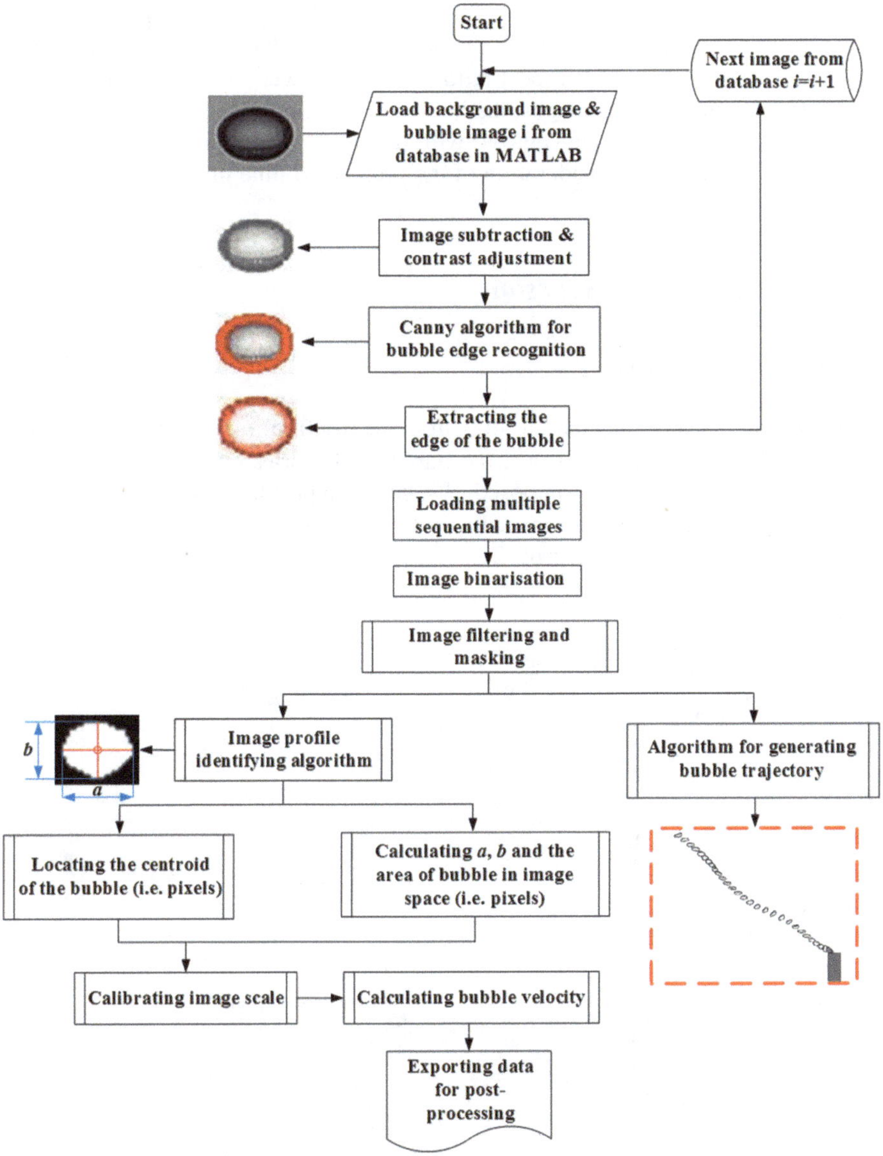

Fig. 4.22 Framework of the bubble image processing code

and arranged sequentially in one single diagram. Therefore, the trajectory of the bubble is obtained.

Based on the observation, most bubbles have an ellipsoidal shape. Therefore, with the code developed, the long axis, *a*, the short axis, *b*, and the centroid position of the bubble are calculated for each bubble. Then the correspondence between pixels in the image and the physical dimensions is established using the image of a ruler. Consequently, the area covered by the bubble, A, the bubble equivalent diameter, d, $d = (4A/\pi)^{0.5}$, and the bubble aspect ratio, E, $E = b/a$, were obtained. For the same bubble in two neighboring images, the bubble displacement is calculated according to the relative position of one centroid with respect to the other; hence, the bubble velocity is calculated in consideration of the predefined time interval.

4.2.2 Results and Discussion

4.2.2.1 Bubble Trajectory

For three flow rates of injected air, 400, 700 and 1000 mL/h, and three water flow velocities, 0.90, 0.97 and 1.03 m/s, the bubble trajectories obtained are plotted in Fig. 4.23. Each subfigure is composed of successive profiles of the same bubble at different moments. As can be seen, for u_l = 0.90 m/s, bubble size is fairly large; moreover, bubbles are featured by flat ellipsoidal shapes. With the increase in the

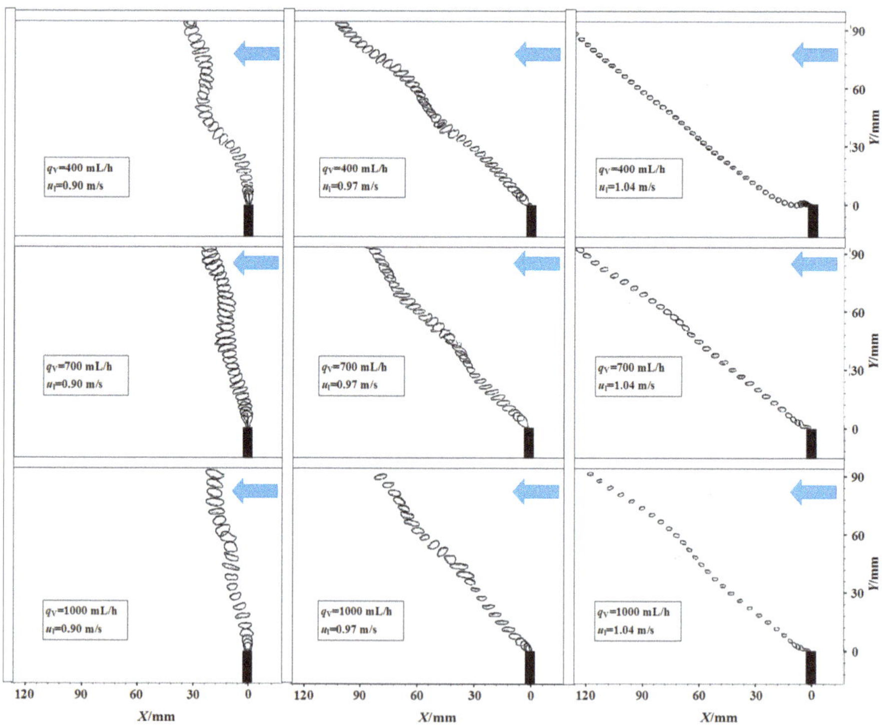

Fig. 4.23 Variation of bubble trajectory pattern with air flow rate and water velocity

water velocity, the bubble size decreases continuously; meanwhile, bubbles tend to be rounded. For u_l = 1.03 m/s, the bubble trajectory deviates remarkably from the vertical direction. In contrast, at u_l = 0.90 m/s, the inclination of the bubble trajectory towards the water flow direction is relatively alleviated; however, the distortion of the bubble trajectory is perceivable, which is particular obvious at q_V = 400 mL/h. In comparison, for u_l = 0.97 m/s and 1.03 m/s, the bubble trajectory is less sensitive to the variation in q_V, as demonstrated in Fig. 4.23.

It is reported that the trajectory of the rising bubble in stagnant liquid incorporates two segments, namely an initial linear segment and then a sinusoidal segment [13]. As can be seen in Fig. 4.23, the situation at u_l = 0.90 m/s and q_V = 400 mL/h is rather similar to that in stagnant liquid. Moreover, it is revealed that the vertical rising of the bubble can maintain for a long distance for low water velocities [63]. In the present study, with the increase in the air flow rate, the initial vertical trajectory segment is apparently shortened. As the water velocity increases, even the trajectory oscillation at the latter stage is attenuated and the bubble trajectory is apparently straightened.

The evolution of the bubble at the initial stage was evidently related to the water velocity. Therefore, two parameters, namely the deviation angle, θ, and the departure distance, s, are defined to further explain the deformation of the bubble trajectory near the nozzle, as illustrated in Fig. 4.24. For q_V = 400 mL/h, the variations in θ and s, which are non-dimensionalized with the bubble equivalent diameter, d, as the function of the water velocity are plotted in Fig. 4.24a, b, respectively. As can be seen in Fig. 4.24a, for u_l < 0.90 m/s, the increase in u_l only leads to a slight increase in the deviation angle. While for u_l > 0.90 m/s, the deviation angle increases sharply with the increase in the water velocity; furthermore, the data points can be approximated with a quadratic relationship, as demonstrated in Fig. 4.24a as well. In this context, it should be pointed out that the bubble trajectory is influenced by not just the water velocity but also the equivalent bubble diameter.

The relationship between s and u_l is similar to that indicated in Fig. 4.24a. In the same fashion, a quadratic relationship is constructed to cover the data points at u_l > 0.90 m/s. The increase in u_l results in a synchronous increase in the shear force exerted on the bubble surface, which is produced due to the velocity difference between the water and the bubble. In this case, the vertical rising of the bubble is subjected to a large resistance. Moreover, it is seen that at high water velocity, large bubble cannot take into shape at the nozzle outlet; alternatively, a small bubble is created as a response to the horizontal disturbance by the water flow. Similar results of bubble deformation have been reported in the work on the evolution of the bubble in the turbulent boundary layer [64]. It is conjectured that as the water velocity increases further, the integrity of the injected air will be further undermined; then the size of the released bubble might be reduced further.

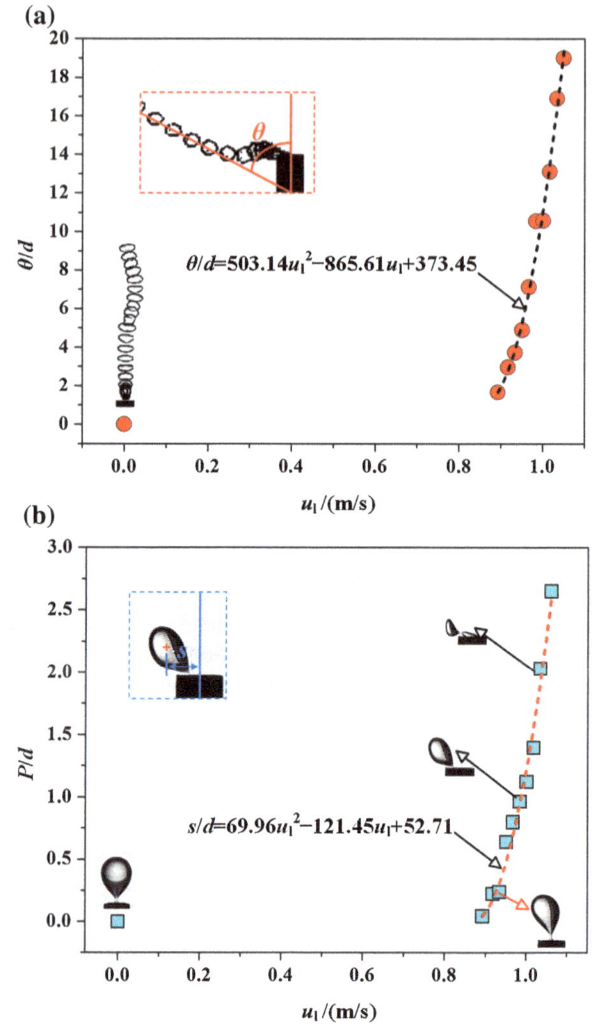

Fig. 4.24 Variation of bubble trajectory parameters with water velocity for q_V = 400 mL/h

4.2.2.2 Bubble Velocity Components

For q_V = 400 mL/h, variations in the bubble velocity components, namely u_b and v_b, as a function of the vertical distance from the nozzle outlet are plotted in Fig. 4.25a, b, respectively. As can be seen, higher water velocity is associated with higher bubble velocity, as is shared by both u_b and v_b. In Fig. 4.25a, for u_l = 1.04 m/s, u_b manifests an abrupt fluctuation as the bubble detaches from the nozzle, as occurs as well in Fig. 4.25b. In this context, the bubble bounds back as it separates entirely from the nozzle. For u_l = 0.90 m/s, negative values of v_b emerge, which is caused by the strong downward suppression of the bubble rising by the water flow. In Fig. 4.25b, for $u_l = 0.97$ m/s, the vertical bubble velocity, v_b, increases sharply to its maximum and then v_b fluctuates around 0.25 m/s. The increase in the water velocity leads to the attenuation of the fluctuation of v_b. In comparison, the discrepancy in horizontal bubble velocity, u_b, between the two cases is remarkable. Situations of low water flow velocities are featured by drastic fluctuations of u_b. Regarding the situation with stagnant water, it has been concluded that the wake vortices generated behind the bubble have a strong influence on the bubble velocity [34]. In the present study, the wake effect is anticipated as well; but such an effect is more evident at lower water velocity. With the increase in the water velocity, the impetus behind the bubble motion in streamwise direction excels the disturbance of wake vortices on the bubble; therefore, variations in both the two bubble velocity components are restricted.

Fig. 4.25 Bubble velocity variation at different water flow velocities

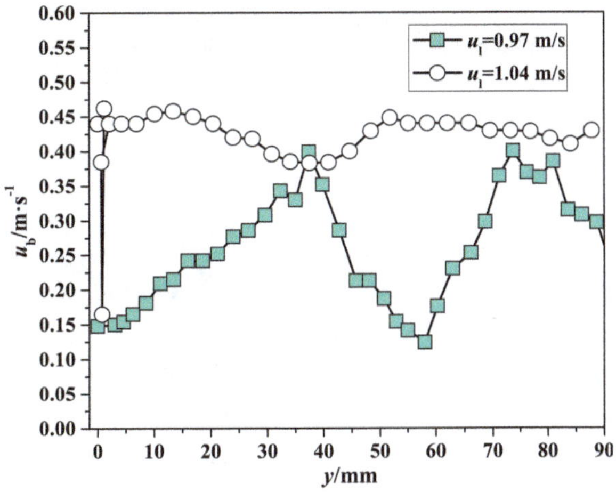

(a) Horizontal velocity component

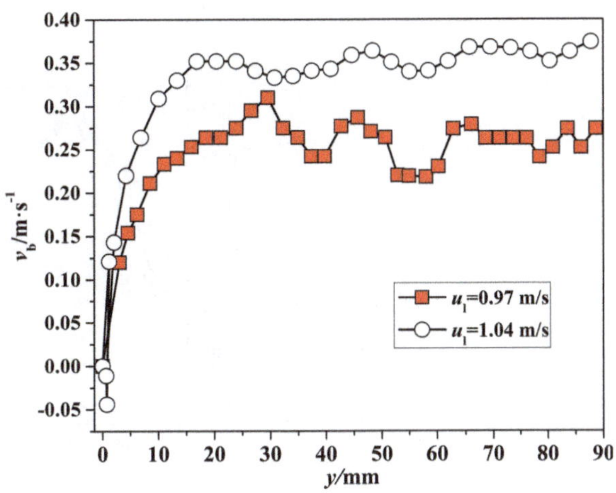

(b) Vertical velocity component

For $u_l = 0.90$ m/s, the variations in u_b and v_b with the movement of the bubble are shown in Fig. 4.26. On the whole, the influence of the air flow rate on the two bubble velocity components is not significant. In Fig. 4.26a, low air flow rate is matched with relatively high streamwise bubble velocity, as is distinct for $Y = 20$–50 mm. Nevertheless, for the vertical velocity component, v_b, the tendency is reversed. Similar conclusions have been obtained with the stagnant liquid [65, 66]. As q_V increases, the increase in bubble size leads to an increase in the buoyant force; therefore, the acceleration stage of the bubble is extended, as indicated in Fig. 4.26b. After a short distance which is dominated by bubble acceleration, v_b fluctuates for a vertical distance and then the bubble terminal velocity is attained at approximately $Y = 45$ mm. In this context, at any given water velocity, the bubble terminal velocity varies only slightly with the increase in the air flow rate.

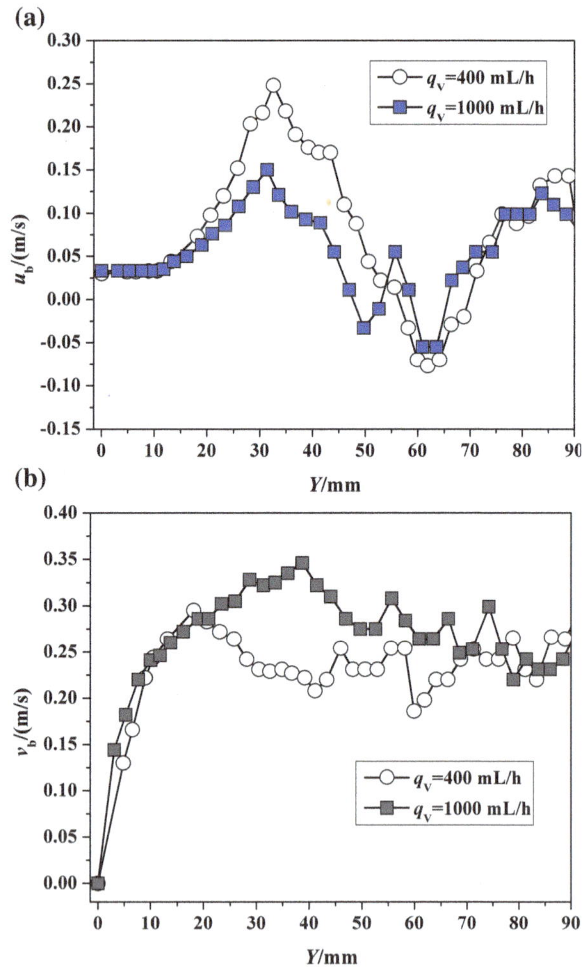

Fig. 4.26 Bubble velocity variations at different air flow rates

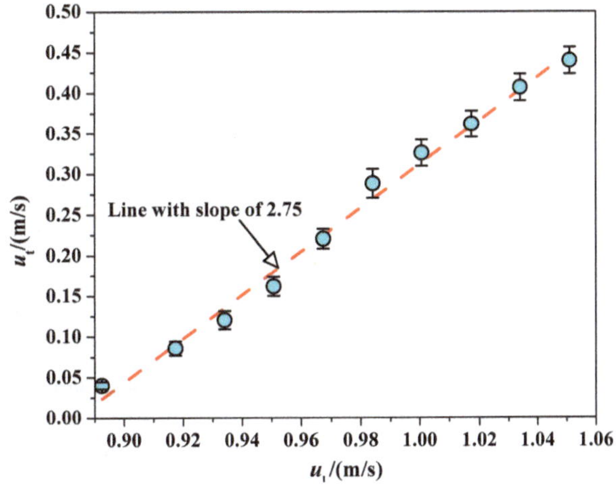

Fig. 4.27 Variation of the horizontal bubble terminal velocity with the water velocity

As can be seen from Figs. 4.25 and 4.26, there is a large velocity gap between the streamwise bubble velocity and the water velocity. To seek the correlation between the two velocities, the streamwise bubble velocity is averaged as the bubble velocity attains the terminal velocity. The variation of the horizontal bubble terminal velocity, u_t, obtained through averaging bubble velocity within 45 mm < Y <90 mm, with the water velocity, u_l, is plotted in Fig. 4.27.

As can be seen, there is a nearly linear relationship between the bubble velocity and the water velocity. A fitting line with a slope of 2.75 is plotted in Fig. 4.27 as well. With the variation in the water velocity, the motion of the bubble will reach a balance state in streamwise direction, as is determined by the impetus from the water and the resistance of the bubble to that force. Furthermore, such a state is related to bubble size as well.

4.2.2.3 Equivalent Bubble Diameter and Bubble Shape

The influence of the air flow rate on the bubble size has been demonstrated in [67]. Here, an extension based on the previous result is attained with the influence of the water velocity being considered. The equivalent bubble diameter, d, is calculated as the water velocity, u_l, is varied from 0.90 to 1.05 m/s. For each velocity, the bubbles do not overlap with each other and individual bubbles are identifiable. The bubble terminal velocity component, u_t, is obtained using the above approach. The correspondence between u_t and d is shown in Fig. 4.28. Meanwhile, the results based on the empirical formula proposed for the stagnant liquid in [68] are presented in Fig. 4.28 as well.

As can be seen, the effects of the water velocity on the bubble terminal velocity are significant. For low water velocity, the horizontal disturbance on the bubble is slight and the data obtained in [68] are predominant. With the increase in the water

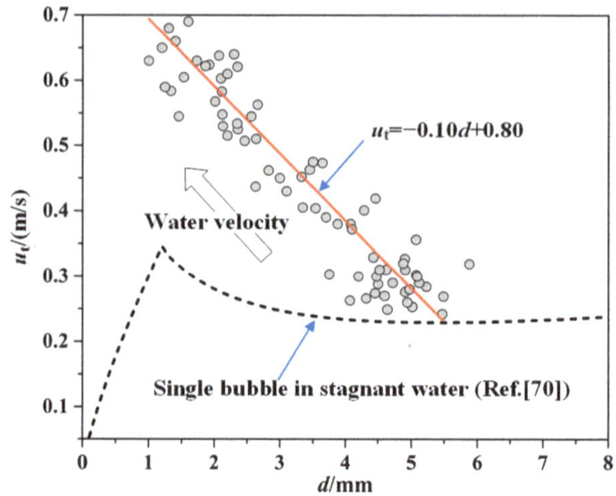

Fig. 4.28 Variation of bubble terminal velocity with the bubble equivalent diameter

velocity, the bubble terminal velocity increases, the equivalent bubble diameter decreases. The sparsely distributed data points can be correlated with a linear relationship, which is illustrated in Fig. 4.28. In this context, although bubbles manifest various size and shape, the geometric and kinematic parameters are inherently related.

Of interest is the bubble deformation in the liquid cross flow, which is described through the bubble aspect ratio, E. The correspondence between E and d is illustrated in Fig. 4.29. The bubble aspect ratio is calculated statistically for each water velocity. As can be seen, there is no explicit relationship between E and d. Nevertheless, three zones are roughly recognizable in Fig. 4.29. Zone A is

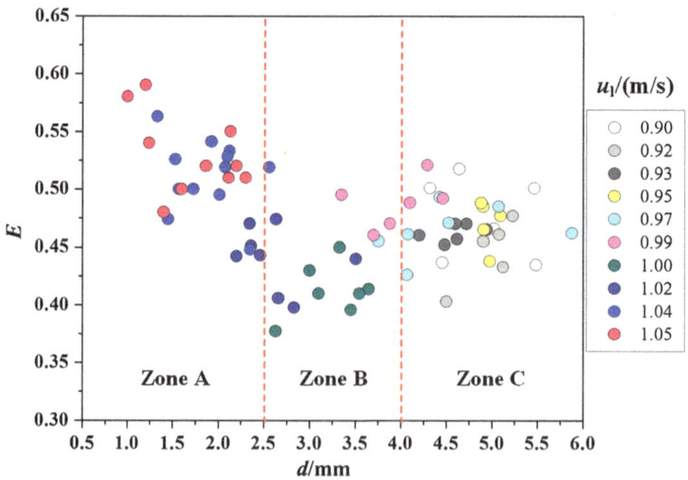

Fig. 4.29 Correspondence between the bubble aspect ratio and the equivalent bubble diameter

associated with high water velocity. In this zone, E is high and d is less than 2.5 mm. In Zone B, the water flow velocity falls within an intermediate range. The values of E indicate that bubbles in Zone B exhibit clearly flat shape; meanwhile, d ranges from 2.5 to 4.0 mm. Regarding Zone C, relatively low water flow velocities are covered; moreover, the equivalent bubble diameter, d, exceeds 4.0 mm and E is relatively low. On the whole, the bubble aspect ratio shown in Fig. 4.29 is lower than 0.6, which is much lower than its counterpart related to stagnant water [69]. Therefore, the effect of the water velocity on the bubble aspect ratio is demonstrated.

4.2.3 Bubble Deformation and Dimensionless Numbers

Regarding the bubbles involved in the liquid cross flow, both the geometric and kinetic features can be traced to the forces exerted on the bubble. Among the forces, the buoyant force and the surface tension deserve particular consideration. The comparison of the forces has usually been implemented through non-dimensional numbers [70]. In this context, the Eötvös number (*Eo*) is defined as:

$$Eo = \frac{g(\rho_l - \rho_g)d^2}{\sigma} \tag{4.20}$$

where ρ_l and ρ_g denote the water and air density, respectively; d is the equivalent bubble diameter, which serves as a characteristic length; σ denotes the surface tension of water.

Essentially, *Eo* is a ratio of the buoyant force to the surface tension force. For different water velocities, the bubble aspect ratio is plotted as a function of *Eo* in

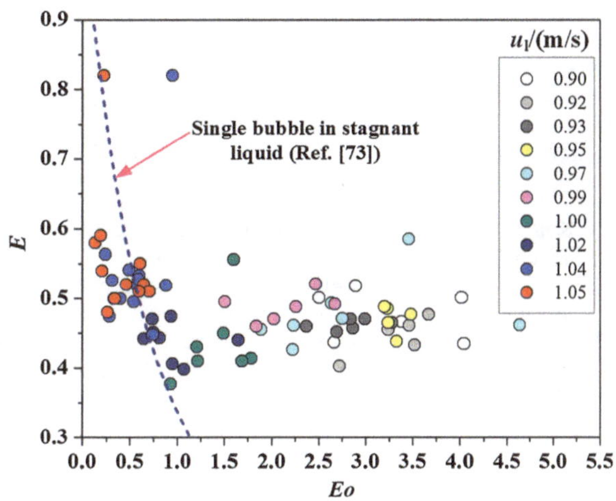

Fig. 4.30 Variation of the bubble aspect ratio with *Eo* for different water velocities

Fig. 4.30. Meanwhile, the relationship between E and Eo for stagnant water proposed in [71] is presented. For $Eo < 1$, the data points of E obtained here are distributed on both sides of the curve obtained in [71]; in this case, the equivalent bubble diameter is less than 2.8 mm, and the water velocity is relative high. Meanwhile, for $Eo < 1$, the bubble aspect ratio is sensitive to the variation in Eo, as can be seen in Fig. 4.30. In contrast, for $Eo > 1$, with the increase in Eo, the bubble aspect ratio fluctuates roughly between 0.4 and 0.6. It should be noted that the Eo values obtained here exceed those associated with stagnant liquid. On the whole, with the increase in u_l, Eo decreases, which implies that the influence of the buoyant force on the bubble decays rapidly relative to that of the surface tension force.

4.2.4 Summary

(1) As the liquid cross flow velocity increases, the deviation of the bubble trajectory relative to the air injection direction becomes remarkable; meanwhile, bubble size decreases and bubble shape shifts from flat ellipsoidal shape to rounded shape. For a given liquid cross flow velocity, the variation in the air flow rate causes a slight change in bubble geometry.
(2) The traveling bubble in the liquid cross flow experiences velocity fluctuations. Velocity fluctuations in both the water flow direction and the air injection direction are intensified as the water velocity decreases. As the balance state for the bubble is attained, there is an approximately linear relationship between the bubble velocity and the water flow velocity.
(3) As the equivalent bubble diameter decreases, the increase in the bubble terminal velocity follows a nearly linear tendency. The bubble terminal velocity in water flow direction is higher than that in stagnant water. The bubble aspect ratio for small Eötvös number agrees with that in the stagnant water; however, large Eötvös number obtained here is beyond the scope related to the stagnant water, and corresponding bubble aspect ratio is low.

References

1. Srinoi P, Sato K. Controllable air-bubbles size generator performance with swirl flow. Eng Technol Appl Sci Res. 2015;8(4):245–249.
2. Onoe K, Wada Y, Matsumoto M. Fascination and engineering application of reaction field utilizing fine bubbles. Jpn J Multiph Flow. 2016;30(1):27–36.
3. Nagami Y, Saito T. An experimental study of the modulation of the bubble motion by gas–liquid-phase Interaction in oscillating-grid decaying turbulence. Flow Turbul Combust. 2014;92(1):147–174.

4. Mikaeliana D, Larcya A, Cockxb A, Wylocka C, Haut B. Dynamics and morphology of single ellipsoidal bubbles in liquids. Exp Thermal Fluid Sci. 2015;64:1–12.
5. Hua J. CFD simulations of the effects of small dispersed bubbles on the rising of a single large bubble in 2D vertical channels. Chem Eng Sci. 2015;123:99–115.
6. Böhm L, Brehmer M, Kraume M. Comparison of the single bubble ascent in a newtonian and a non-newtonian liquid: a phenomenological PIV study. Chem Ing Tec. 2016;88(1–2): 93–106.
7. Valiorgue P, Souzy N, El Hajem M, Hadid HB, Simoëns S. Concentration measurement in the wake of a free rising bubble using planar laser-induced fluorescence (PLIF) with a calibration taking into account fluorescence extinction variations. Exp Fluids. 2014;55(4):1–2.
8. Huang J, Saito T. Influence of bubble-surface contamination on instantaneous mass transfer. Chem Eng Technol. 2015;38(11):1947–1954.
9. Yoshida K, Morioka S, Kagawa Y, Koyama D, Watanabe Y. Power-law dependence describing subharmonic generation from a non-spherically oscillating bubble. Acoust Sci Technol. 2015;36(3):191–200.
10. Dehaeck S, van Beeck JPAJ, Riethmuller ML. Extended glare point velocimetry and sizing for bubbly flows. Exp Fluids. 2005;39(2):407–419.
11. Akhmetbekov YK, Alekseenko SV, Dulin VM, Markovich DM, Pervunin KS. Planar fluorescence for round bubble imaging and its application for the study of an axisymmetric two-phase jet. Exp Fluids. 2010;48(4):615–629.
12. Wu MM, Gharib M. Experimental studies on the shape and path of small air bubbles rising in clean water. Phys Fluids. 2002;14(7):L49–52.
13. Liu L, Yan H, Zhao G. Experimental studies on the shape and motion of air bubbles in viscous liquids. Exp Thermal Fluid Sci. 2015;62:109–121.
14. Brücker C. Structure and dynamics of the wake of bubbles and its relevance for bubble interaction. Phys Fluids. 1999;11(7):1781–1796.
15. Cano-Lozano JC, Bohorquez P, Martínez-Bazán C. Wake instability of a fixed axisymmetric bubble of realistic shape. Int J Multiph Flow. 2013;51(51):11–21.
16. Zawala J, Wiertel A, Niecikowska A, Małysa K. Influence of external vibrations on bubble coalescence time at water and oil surfaces-Experiments and modelling. Colloids Surf A Physicochem Eng Aspects. 2017;519:137–145.
17. Maldonado M, Quinn J, Gómez CO, Finch JA. An experimental study examining the relationship between bubble shape and rise velocity. Chem Eng Sci. 2013;98(29):7–11.
18. Ortiz-Villafuerte J, Schmidl WD, Hassan YA. Three-dimensional PIV measurements of bubble drag and lift coefficients in restricted media. Rev Mex Fís. 2013;59:444–452.
19. Quinn J, Maldonado M, Gómez CO, Finch JA. Experimental study on the shape–velocity relationship of an ellipsoidal bubble in inorganic salt solutions. Miner Eng. 2014;55(1):5–10.
20. Wang D, Zhang X, Wang J, Koide T. Effect of small amount of water on CO_2 bubble behavior in ionic liquid systems. Ind Eng Chem Res. 2014;53(1):428–439.
21. Okawa T, Tanaka T, Kataoka I, Mori M. Temperature effect on single bubble rise characteristics in stagnant distilled water. Int J Heat Mass Transf. 2003;46(5):903–913.
22. Laqua K, Malone K, Hoffmann M, Krause D, Schlüter M. Methane bubble rise velocities under deep-sea conditions-influence of initial shape deformation. Colloids Surf A Physicochem Eng Aspects. 2016;505:106–117.
23. Celata GP, Cumo M, D'Annibale F, Di Marco P, Tomiyama A, Zovini C. Effect of gas injection mode and purity of liquid on bubble rising in two-component systems. Exp Thermal Fluid Sci. 2006;31(1):37–53.
24. Mulgrew B, Grant PJ, Thompson J. Digital signal processing: concepts and applications. Palgrave Macmillan; 2002.
25. Rahnama M, Gloaguen R. TecLines: a MATLAB-based toolbox for tectonic lineament analysis from satellite images and DEMs, part 2: line segments linking and merging. Remote Sens. 2014;6(11):11468–11493.
26. Meher SK. Recursive and noise-exclusive fuzzy switching median filter for impulse noise reduction. Eng Appl Artif Intell. 2014;30(3):145–154.

27. Ding L, Goshtasby A. On the Canny edge detector. Pattern Recogn. 2001;34(3):721–725.
28. Bongiovanni C, Chevaillier JP, Fabre J. Sizing of bubbles by incoherent imaging: defocus bias. Exp Fluids. 1997;23:209–216.
29. Dehaeck S, Van Parys H, Hubin A, van Beeck JPAJ. Laser marked shadowgraphy: a novel optical planar technique for the study of microbubbles and droplets. Exp Fluids. 2009;47:333–341.
30. Celata GP, D'Annibale F, Di Marco P, Memoli G, Tomiyama A. Measurements of rising velocity of a small bubble in a stagnant fluid in one- and two-component systems. Exp Thermal Fluid Sci. 2007;31(6):609–623.
31. Zhang AM, Cui P, Cui J, Wang QX. Experimental study on bubble dynamics subject to buoyancy. J Fluid Mech. 2015;776:137–160.
32. Zhang AM, Cui P, Wang Y. Experiments on bubble dynamics between a free surface and a rigid wall. Exp Fluids. 2013;54(10):1–18.
33. Dukhin SS, Kovalchuk V, Gochev G, Lotfi M, Krzan M, Malysa K, Miller R. Dynamics of rear stagnant cap formation at the surface of spherical bubbles rising in surfactant solutions at large Reynolds numbers under conditions of small Marangoni number and slow sorption kinetics. Adv Colloid Interface Sci. 2015;222:260–274.
34. Ellingsen K, Risso F. On the rise of an ellipsoidal bubble in water: oscillatory paths and liquid-induced velocity. J Fluid Mech. 2001;440:235–268.
35. Das D. Study of air bubble injection into laminar and turbulent water flow through an annular test section. CUNY City College; 2012.
36. Tomiyama A, Kataoka I, Zun I, Sakaguchi T. Drag coefficients of single bubbles under normal and micro gravity conditions. JSME Int J Ser B Fluids Thermal Eng. 1998;41(2): 472–489.
37. Mendelson HD. The prediction of bubble terminal velocities from wave theory. AIChE J. 2010;13(2):250–253.
38. Tomiyama A, Celata GP, Hosokawa S, Yoshida S. Terminal velocity of single bubbles in surface tension force dominant regime. Int J Multiph Flow. 2002;28(9):1497–1519.
39. Mikaelian D, Larcy A, Dehaeck S, Haut B. A new experimental method to analyze the dynamics and the morphology of bubbles in liquids: application to single ellipsoidal bubbles. Chem Eng Sci. 2013;100(3):529–538.
40. Aoyama S, Hayashi K, Hosokawa S, Tomiyama A. Shapes of ellipsoidal bubbles in infinite stagnant liquids. Int J Multiph Flow. 2016;79(15):23–30.
41. Moore DW. The velocity of rise of distorted gas bubbles in a liquid of small viscosity. J Fluid Mech. 1965;23(4):749–766.
42. Taylor TD, Acrivos A. On the deformation and drag of a falling viscous drop at low Reynolds number. J Fluid Mech. 1964;18(3):466–476.
43. Sugihara K, Sanada T, Shirota M, et al. Behavior of single rising bubbles in superpurified water. Kagaku Kogaku Ronbunshu. 2007;33(5):402–408.
44. Tadaki T, Maeda S. On the shape and velocity of single air bubbles rising in various liquids. Soc Chem Eng. 1961;25(4):254–264.
45. Myint W, Hosokawa S, Tomiyama A. Shapes of single drops rising through stagnant liquids. J Fluid Sci Technol. 2007;2(2):184–195.
46. Grace JR, Wairegi T, Nguyen TH. Shapes and velocities of single drops and bubbles moving freely through immiscible liquids. Chem Eng Res Des. 1976;54(3):167–173.
47. Navisa J, Sravya T, Swetha M, Venkatesan M. Effect of bubble size on aeration process. Asian J Sci Res. 2014;7(4):482–487.
48. Müller-Fischer N, Tobler P, Dressler M, Fischer P, Windhab EJ. Single bubble deformation and breakup in simple shear flow. Exp Fluids. 2008;45:917–926.
49. Yamaguchi T, Iguchi M, Uemura T. Behavior of a small single bubble rising in a rotating flow field. Exp Mech. 2004;44:533–540.
50. Balzán MA, Sanders RS, Fleck BA. Bubble formation regimes during gas injection into a liquid cross flow in a conduit. Can J Chem Eng. 2017;95:372–385.

51. Ravelet F, Colin C, Risso F. On the dynamics and breakup of a bubble rising in a turbulent flow. Phys Fluids. 2011;23:103301.
52. Chen XP, Shao DC. Measuring bubble size in aerated flow. J Hydrodyn Ser B. 2006;18: 470–474.
53. Yapa PD, Dasanayaka LK, Bandara UC, Nakata K. A model to simulate the transport and fate of gas and hydrates released in deepwater. J Hydraul Res. 2010;48(5):559–572.
54. Xu Y, Ersson M, Jönsson PG. A Mathematical modeling study of bubble formations in a molten steel bath. Metall Mater Trans B. 2015;46B:2628–2638.
55. Zhang W, Zhu DZ. Trajectories of air-water bubbly jets in crossflows. J Hydraul Eng. 2014;140:06014011.
56. Liu C, Liang B, Tang S, Zhang H, Min E. A theoretical model for the size prediction of single bubbles formed under liquid cross-flow. Chin J Chem Eng. 2010;18:770–776.
57. Zhang W, Zhu DZ. Bubble characteristics of air–water bubbly jets in crossflow. Int J Multiph Flow. 2013;55:156–171.
58. Kang C, Zhang W, Yiping G, Mao N. Bubble size and flow characteristics of bubbly flow downstream of a ventilated cylinder. Chem Eng Res Des. 2017;122:263–272.
59. Martínez-bazán C, Montañés JL, Lasheras JC. On the breakup of an air bubble injected into a fully developed turbulent flow. Part 1. Breakup frequency. J Fluid Mech. 1999;401:157–182.
60. Baylar A, Ozkan F, Unsal M. Effect of air inlet hole diameter of venturi tube on air injection rate. KSCE J Civ Eng. 2010;14:489–492.
61. Wei W, Deng J, Zhang F, Tian Z. A numerical model for air concentration distribution in self-aerated open channel flows. J Hydrodyn Ser B. 2015;27:394–402.
62. Ding L, Goshtasby A. On the Canny edge detector. Pattern Recogn. 2001;34:721–725.
63. Jobehdar MH, Gadallah AH, Siddiqui K, Chishty WA. Investigation of the bubble formation in liquid cross-flow using a novel nozzle design. In: Proceedings of the ASME 2013 fluids engineering division summer meeting; 2013 July 7–11, Nevada, USA.
64. Katoh K, Arii Y, Wakimoto T. Bubble formation from an air jet injected into a turbulent boundary layer. J Fluid Sci Technol. 2011;6(4):528–541.
65. Zhang X, Dong H, Bao D, Huang Y, Zhang X, Zhang S. Effect of small amount of water on CO_2 bubble behavior in ionic liquid systems. Ind Eng Chem Res. 2014;53:428–439.
66. Wang B, Socolofsky SA. On the bubble rise velocity of a continually released bubble chain in still water and with crossflow. Phys Fluids. 2015;27:103301.
67. Ghaemi S, Rahimi P, Nobes DS. The effect of gas-injector location on bubble formation in liquid cross flow. Phys Fluids. 2010;22:043305.
68. Tomiyama A, Celata GP, Hosokawa S, Yoshida S. Terminal velocity of single bubbles in surface tension force dominant regime. Int J Multiph Flow. 2002;28:1497–1519.
69. Mikaelian D, Larcy A, Dehaeck S, Haut B. A new experimental method to analyze the dynamics and the morphology of bubbles in liquids: application to single ellipsoidal bubbles. Chem Eng Sci. 2013;100:529–538.
70. Dong H, Wang X, Liu L, Zhang X, Zhang S. The rise and deformation of a single bubble in ionic liquids. Chem Eng Sci. 2010;65:3240–3248.
71. Okawa T, Tanaka T, Kataoka I. Temperature effect on single bubble rise characteristics in stagnant distilled water. Int J Heat Mass Transf. 2003;46:903–913.

Chapter 5
Wake Flow of the Ventilation Cylinder

Abstract Injecting air into water serves as an effective measure of generating bubbles. The characteristics of the background water flow must exert a significant effect on the bubbles trapped. In this chapter, it is intended to reveal the bubbly flow pattern downstream of a cylinder. Therefore, a cylinder is installed in the horizontal transparent section of a water tunnel. Particle image velocimetry technique is used to measure flow velocity downstream of the cylinder under no-ventilation condition. Air is injected into water flows through a flow passage inside the cylinder. High-speed photography, in association with a LED light source, is utilized to capture consecutive images of moving bubbles downstream of the cylinder. With the research techniques, it is expected to obtain primary bubble parameters such as velocity, Sauter mean diameter and volume fraction. Furthermore, the relationship between carrier flow characteristics and bubbly flow pattern is expected to be established.

5.1 Experimental Setup

For bubbly flows, the relationship between bubbly flow pattern and the carrier flow characteristics is of prime importance in consideration of the bubble movement, deformation and collapse. Even with adequate carrier flow information, modeling the bubble motion process is not an easy task, since that turbulent fluctuations residing in the carrier flow exert a complicated effect on bubbles. An overview of the treatment of such an issue has been presented in [1]. Until now, bubbly flows in pipes and columns, which share simple boundary conditions, have been studied extensively. In contrast, few studies of bubbly flows around a blunt body have been reported.

Recently, a promising experimental technique, namely shadow image velocimetry (SIV), is attempted to trace bubbles [2]. The central parts of such a technique lie in two aspects, high-speed photography and image processing. Consecutive bubble images are captured consecutively. Hence, bubbles in the captured images, overlapped or separated, are identified and statistically manipulated using the specifically developed algorithm.

C. Kang et al., *Methods for Solving Complex Problems in Fluids Engineering*,
https://doi.org/10.1007/978-981-13-2649-3_5

5.1.1 Water Tunnel and Ventilated Cylinder

The experiment is carried out in the water tunnel schematically shown in Fig. 4.17. The dimensions of the horizontally mounted test section are 700 mm × 50 mm× 315 mm. All the four sides of the test section are transparent. A circular cylinder with the diameter and length of 30 and 49 mm, respectively, is installed in the test section. Mounted upstream of the test section, an elaborately designed settling chamber produces a homogeneous upstream water flow. Meanwhile, turbulent intensity of the upstream water flow can be lowered down to a certain level.

Primary components of a ventilation system are shown in Fig. 5.1. The stabilization tank can provide fairly stable air pressure. The air flow rate is read from the rotameter. There is one hole on the cylinder surface for discharging air into the water tunnel, and the hole is positioned at the center of the cylinder in spanwise direction. Such a scheme ensures a two-dimensional flow on the cross section vertical to the cylinder axis, which serves as the DOV plane, and the observation of the bubbly flow patterns is facilitated.

Injecting gas into a downward liquid pipe flow has been attempted and a cone-shaped cavity with its size comparable to the inner diameter of the pipe is produced through adjusting relevant parameters [3]. In [4], air is injected into the water flow in a water tunnel and supercavity encompassing a submerged cavitator is induced. Here, there is only one hole with the diameter of 1 mm in the cylinder, as shown in Fig. 5.1. The air flow rate can be tuned through the pressure valve;

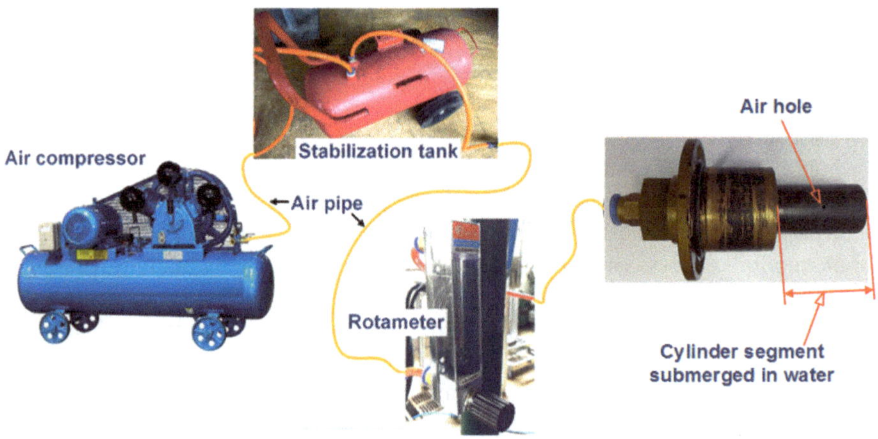

Fig. 5.1 Primary components of the ventilation system

therefore, bubbles are dispersedly distributed and bubble size is small. According to the explanation of the effect of Stokes number made in [5], the collisional Stokes number in the bubbly wake flow is of the magnitude order of 1.

Bubbles transported by the wake flow downstream of a cylinder bear a background of well-acknowledged carrier flow features. There are several factors that are supposed to affect the bubbly flow pattern. Akbar et al. transformed a symmetric bubbly flow pattern to an asymmetric bubbly flow pattern in a rectangular column via two arrays of injecting nozzles [6]. In the present study, the position of the injection hole is fixed and only the air flow rate is adjusted.

5.1.2 Optical Configuration

The pure-water wake flow is measured with a LaVision PIV system, which is used in Chap. 4. A Nd: YAG laser with the light wavelength of 532 nm is used to illuminate the flow regions. A CCD camera of Imager Pro SX 5M, featuring an image resolution of 2456 × 2058 pixels, is used. The maximum image-capturing frequency is 14.5 fps. Hollow glass particles with diameters ranging from 20 to 50 μm are used as tracing particles. Data acquisition and processing are conducted using the Davis software. The Tecplot software is used to process acquired data and to calculate flow quantities.

With the light source being replaced by an LED light source, the SIV technique is utilized to capture bubbly flow images. As shown in Fig. 5.2, the light source and the camera are located at the two sides of the test section. With the mode of double-frame and double-exposure, the time interval between two neighboring frames is 600 ns.

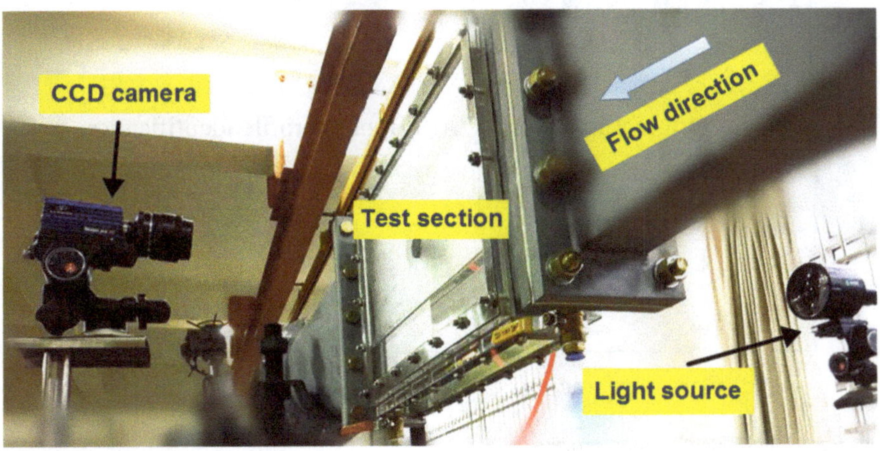

Fig. 5.2 Test section and optical configuration of SIV

During the experiment, after each individual case, the entire loop will be emptied and then refilled with pure water. The disturbance of residual bubbles and dissolve air is thereby avoided. The maximum upstream velocity in the experiment is 7 m/s; thus cavitation will not occur. Apart from the above consideration, the experimental rig is carefully calibrated. The overall uncertainty of the experiment is less than 0.5%, including the uncertainties of the optical instruments.

5.1.3 Bubbles Separation Algorithm

For bubble image processing, it is crucial to separate bubbles from the images captured with high-speed photography. Here, the Canny algorithm embedded in MATLAB software is used as the basic module of an in-house code. Both image processing and bubble identification are realized with this code. The intermediate stages of such a code are illustrated in Fig. 5.3. In Fig. 5.3a, the background image has been subtracted. Then a preliminary profile generation is realized in Fig. 5.3b. Through curvature tracing, it is expected to separate those bubbles overlapping with each other, as indicated in Fig. 5.3c. With the double-frame and double exposure technique, the displacement of the same bubble in two successive images can be calculated with the reference of the movement of the mass center of the bubble. The bubble profiles marked in Fig. 5.3d helps to determine the mass center after the bubble identification. Thereby, the bubble velocity is obtained. All bubbles are assumed to be ellipsoids. Since bubbles in the images are displayed with a planar

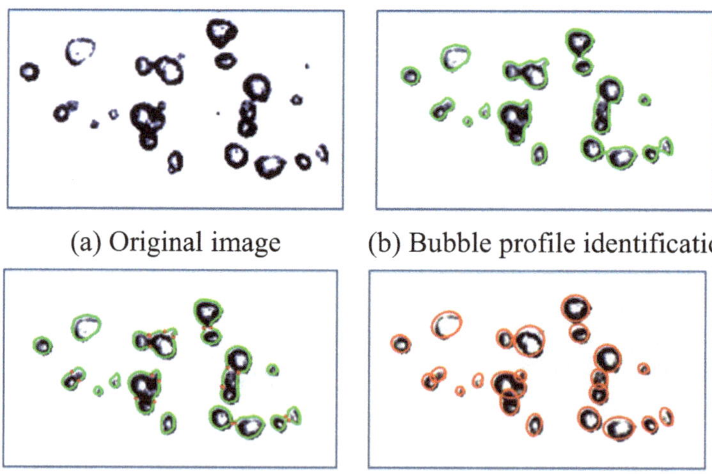

(a) Original image (b) Bubble profile identification

(c) Curvature tracing (d) Separation of individual bubbles

Fig. 5.3 Bubble identification process

view, the dimension in the other direction has to be assumed. Then the volume of individual bubbles can be calculated. This code assumes more functions than that developed in Chap. 4. Here, the separation of overlapped bubbles using the code is particularly emphasized.

5.2 Velocity and Vorticity Distributions

At upstream velocities of 3, 4, 5 and 7 m/s, pure-water wake flows immediately downstream of the cylinder are measured under no-ventilation condition. Cross-sectional distributions of instantaneous vorticity, as well as streamlines, are shown in Fig. 5.4, where the flow direction is from right to left.

A distinct difference between Fig. 5.4a, b lies in the wake width. At low upstream velocity, the wake width is large, accompanied by a large-scale vortex immediately downstream of the cylinder. In contrast, in Fig. 5.4b, the wake is restricted in a rather limited region. Regarding the vorticity magnitude, high upstream velocity is associated with relatively high vorticity magnitude. At U_0 = 3 m/s, the vorticity elements are sparsely distributed in the wake flow. It can be inferred that bubbles entrapped in the wake flows displayed in Fig. 5.4a, b will face the threat of collapse. Furthermore, the oscillation of streamlines is apparent in Fig. 5.4a, while at U_0 = 7 m/s, the streamlines from upper and lower sides rapidly join together after the wake region.

Cross-sectional distributions of averaged velocity and vorticity at the two upstream velocities are shown in Fig. 5.5. In each subfigure, three cross sections are selected and exhibited. Fairly smooth curves are observed for each case. The same tendency is shared by the two subfigures. As the distance from the cylinder increases, velocity deficit drops and the velocity difference between wake center and wake edge diminishes. The most prominent difference between Fig. 5.5a, b is

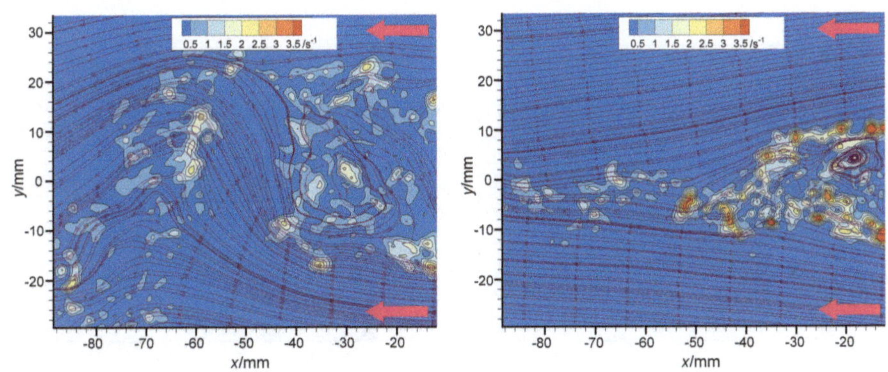

(a) U_0=3 m/s (b) U_0=7 m/s

Fig. 5.4 Instantaneous vorticity distributions and streamline at 3 and 7 m/s (The origin of the coordinates system coincides with the center of the cross section of the cylinder)

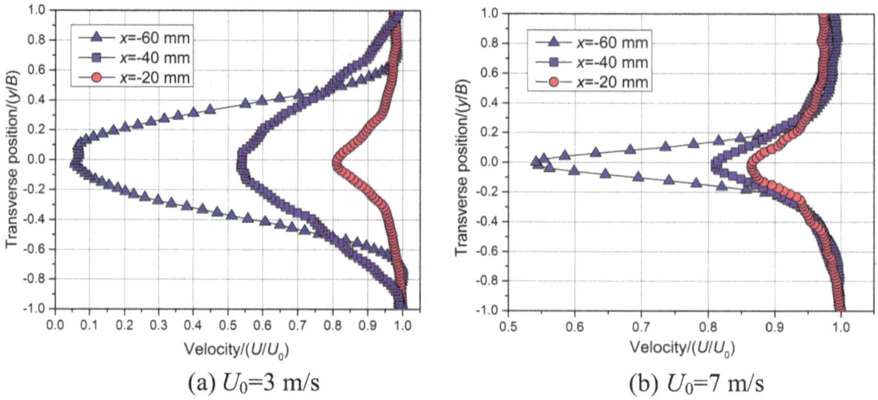

Fig. 5.5 Average velocity distributions at 3 and 7 m/s

the location of high velocity gradients in transverse direction. Literally, this has been reflected in Fig. 5.5, which indicates that different upstream velocities correspond to different wake widths.

In the same manner, averaged vorticity distributions are plotted in Fig. 5.6. Near the cylinder, twin peaks are symmetrically located at the upper and lower sides of the wake centerline. For U_0 = 7 m/s, the two peaks are close to each other. As the wake progresses, the overall vorticity magnitude decreases and cross-sectional vorticity distribution tends to be uniformized. It is noticeable that the normalized transverse position where vorticity peak protrudes varies slightly with x. According to the twin vorticity peaks, the wake flow zone is divided into three distinct subzones along the transverse direction, namely the upper subzone, middle subzone and bottom subzone. The middle subzone is located between the two peaks and contains vortices of various scales, as is related to the wake width.

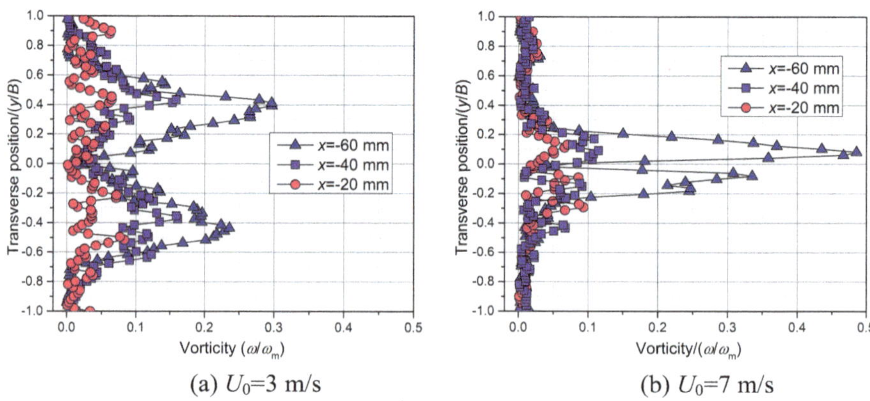

Fig. 5.6 Average vorticity distributions at 3 and 7 m/s

5.3 Bubbles Size Prediction

5.3.1 Bubbly Flow Patterns

In Fig. 5.7, at the same air injection rate q_V of 20 mL/min, temporal bubble flow patterns are exhibited at different upstream velocities to present a whole view of the bubbly wake flow pattern. It is seen that bubbles follow the carrier flow and display a meandering pattern. This is in accordance with transient wake flow pattern downstream of the cylinder. Although an ultimate correspondence between bubble movement and the carrier flow is uncertain, the influence of the carrier flow to bubbly flow pattern is explicit. In this connection, an energy balance between bubbles and the carrier flow is expected, and such a balance applies not just to large-scale flow structures but also to turbulent fluctuations.

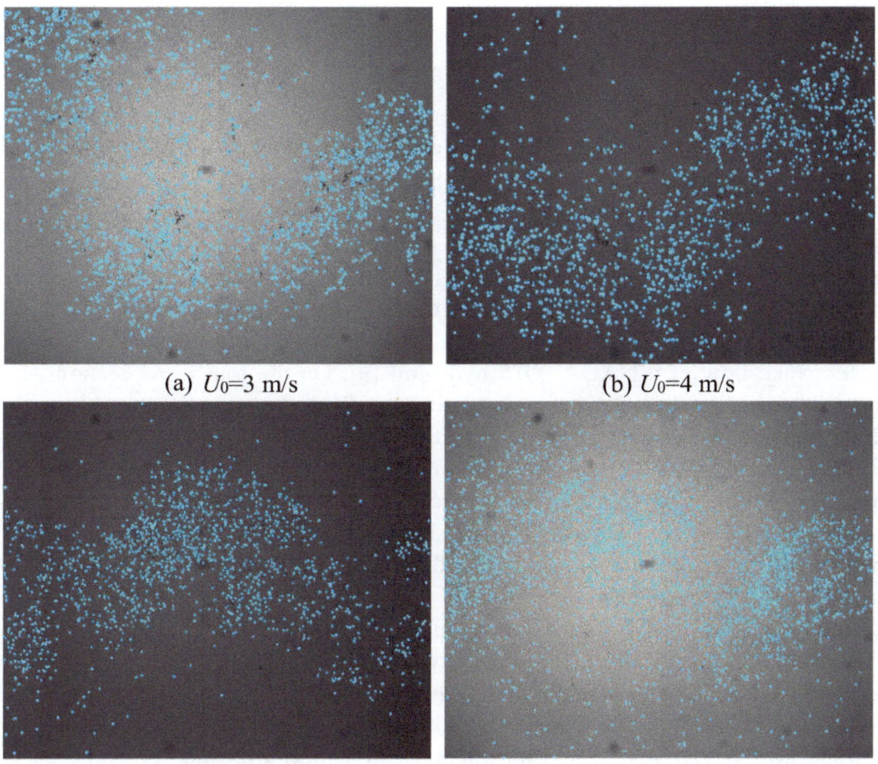

(a) U_0=3 m/s (b) U_0=4 m/s

(c) U_0=5 m/s (d) U_0=7 m/s

Fig. 5.7 Bubble images at different upstream velocities (q_V = 20 mL/min)

5.3.2 Bubble Size

Sauter mean diameter is calculated based on captured bubble images. Here, bubbles are assumed to be ellipsoids, which is in accordance with the suggestion in [7]. Therefore, with respect to each ellipsoidal bubble, the length of the axis in spanwise direction is assumed to be:

$$c = \sqrt{ab} \tag{5.1}$$

where a and b are the lengths of the major and minor axes, respectively. Such an approximation of the three-dimensional bubble profile is reasonable from a statistical viewpoint. Two cameras are employed in [8] to simultaneously capture bubble images from two different view angles and to observe both shape and position of individual bubbles. Such a practice still necessitates statistical analysis based on a sufficiently large sample set.

In consideration of the irregularity of bubble shape, Sauter mean diameter, d_{32}, is used to denote the bubble size and is defined by:

$$d_{32} = 6\frac{V_b}{A_b} \tag{5.2}$$

where V_b and A_b are the volume and surface area of the ellipsoid bubble, respectively. Based on 1024 consecutive images, the cross-sectional distribution of mean Sauter mean diameter at U_0 = 3 m/s is obtained and shown in Fig. 5.8.

As shown in Fig. 5.8, Sauter mean diameter covers a wide spectrum. In the upper and lower subzones, Sauter mean diameter is comparatively small. In the wake center subzone, a uniform bubble size distribution is seen. There is a sharp decrease of bubble size near the interface of the wake center subzone and the upper or the lower subzone. Such a distribution patterns is similar to the results obtained in the wake flow of a NACA 0015 hydrofoil [9]. On the four cross sections, the large Sauter mean diameter is nearly equivalent. With the progression of the wake flow, Sauter mean diameter values tend to be homogenized, as is finally demonstrated on the cross section of x = −60 mm. It is noticeable that on the cross section of x = −20 mm, close to the cylinder, the distribution of Sauter mean diameter is consistent with its counterparts. This cross section suffers from intense interaction between vortices. In this context, the distribution of Sauter mean diameter at U_0 = 3 m/s is representative.

Bubble size is related to turbulent fluctuations. Batchelor suggested in 1951 that there should be a critical bubble size equivalent to a minimum distance over which turbulent kinetic energy budget can attain equilibrium [10]. In 1955, Hinze analyzed the effect of flow characteristics on bubble shape and obtained conclusions at various flow fields [11]. Based on the work of Hinze, Rigby et al. proposed an empirical formula [12]:

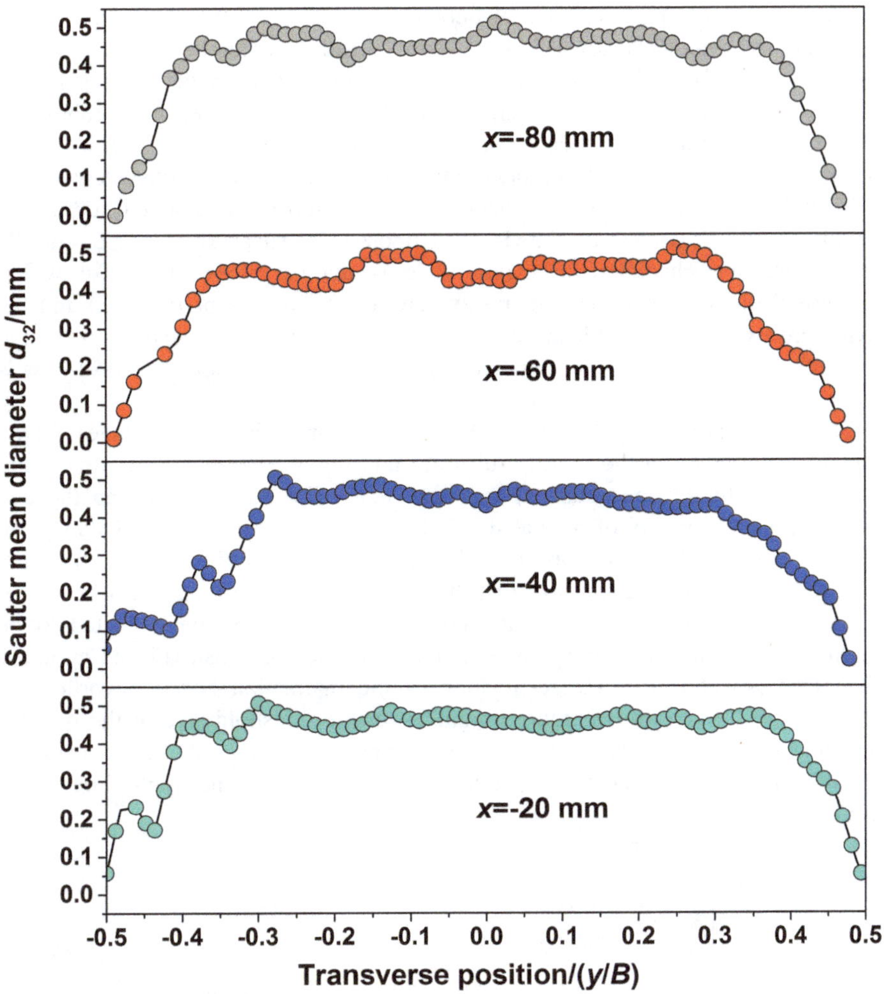

Fig. 5.8 Cross-sectional distribution of Sauter mean diameter

$$d_{32} = C_2 \varepsilon^{-2/5} \left(\frac{We_c \sigma}{2\rho_w} \right)^{3/5} \quad (5.3)$$

where d_{32} is the critical Sauter mean diameter, which indicates the maximum Sauter mean diameter of the bubble that can survive in a certain turbulent environment, C_2 is a constant, ε is the turbulent kinetic energy dissipation rate, We_c is the critical Weber number, σ is the surface tension, and ρ_w is the density of the carrier flow. This formula has been tested under different flow conditions and the most desirable results are seen in homogeneous and isotropic flows. For shear flows, 1.18 and 0.70 are assigned to We_c and C_2, respectively.

The relationship between critical bubble size and turbulent length scale is of particular interest. It is advised in [13] to consider comprehensively turbulent kinetic energy dissipation rate, characteristic length and Reynolds number in isotropic turbulent pipe flows. It is argued in [14] that the existence of bubbles is related to the variation of streamwise integral length scale. In [15, 16], it is confirmed that particle size and turbulent length scale could be connected statistically.

Turbulent kinetic energy dissipation rate, the most fundamental turbulence quantity in Eq. (5.3), is calculated based upon 1024 successively captured PIV images and the characteristic length scale is defined to be the wake width. A comparison of Sauter mean diameter between the results of Eq. (5.3) and the image-processing result is illustrated in Fig. 5.9. Here, a streamwise cross section at $x = -80$ mm serves as a representative, and on this cross section, the wake flow is deemed as fully developed.

The two curves plotted in Fig. 5.9 share similar variation tendencies. Large bubbles accumulate in the middle subzones and bubble size decreases drastically from the middle subzone edge to the wake edge. For the gap between the two curves, it is the imparity of bubble size in the middle subzone, where Sauter mean diameter values obtained from Eq. (5.3) are larger than the image-processing results. Furthermore, bubble size distribution associated with Eq. (5.3) bear dual peaks. In one aspect, values of the constants in Eq. (5.3) are expected to be modified according to characteristics of the wake flow downstream the cylinder. The other aspect is the underestimation of the interaction between vortices of various scales. Such an interaction might lower the threshold size of the bubbles that can survive. Certainly, there might be some other factors such as the change in turbulent fluctuations due to the participation of bubbles, the approximation of surface tension.

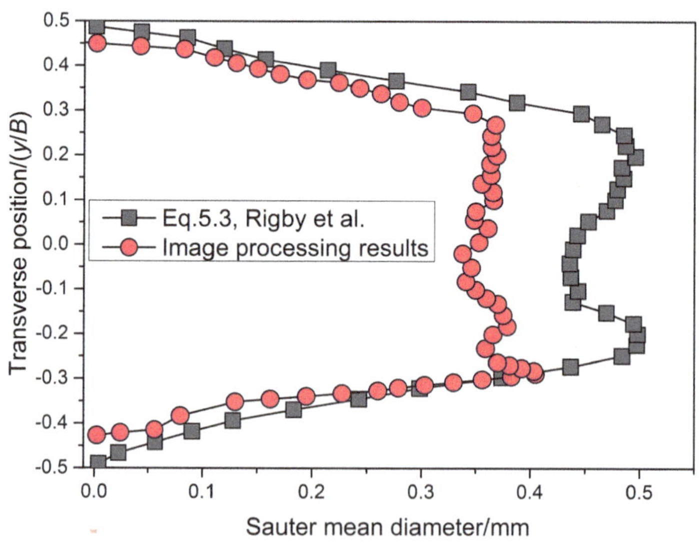

Fig. 5.9 Comparison of bubble size between the present study and Eq. (5.3) at 7 m/s

5.3.3 Comparison of Bubble Size Distribution

The bubble size distribution discussed above is analogous to the cross-sectional bubble size distribution in bubbly pipe flows obtained in [17]. In the pipe flow, uniform bubble size distribution in the middle zone is also proved. And a sharp drop of bubble size is evident near the pipe wall. In the pipe flow, large bubbles near pipe wall are annihilated by the velocity gradients near the pipe wall. By contrast, in the bubbly wake flow, high vorticity near the wake edge promotes the elimination of large bubbles. Furthermore, it is observed that bubble size varies violently near the inner wall of the pipe and it is estimated that bubbles are elongated due to high Reynolds shear stress [18]. When pipes are mounted horizontally, both bubble size and void fraction distributions are appreciably different due to buoyancy effect, as can be justified with the results obtained in [19].

5.4 Bubble Velocity Distribution

At U_0 = 3 m/s, averaged bubble velocity distributions on four cross sections are plotted in Fig. 5.10. Compared with Fig. 5.6a, it is apparent that bubble velocity magnitude is smaller than that of water velocity. In the light of the magnitudes of Reynolds number and air flow rate adopted here, a distinct difference of water velocity distribution between pure water wake and bubbly wake is not anticipated. A distinct feature of Fig. 5.10 is that the middle subzone is dominated by rather

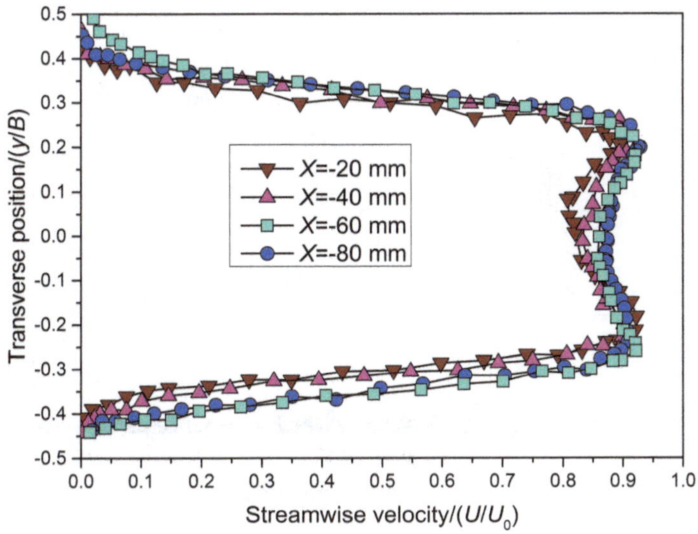

Fig. 5.10 Cross-sectional bubble velocity distributions at 3 m/s

uniform bubble velocity distributions. This is considerably different with Fig. 5.6a. Similarly, for pipe flows, in both [20, 21], it is argued that bubbles could flatten water velocity distribution; meanwhile, both Fujiwara et al. and Aloui et al. prove that bubble velocity in pipe flows also has a rather flattened distribution conforming to the widely accepted power law relation [22, 23]. In the middle subzone, the air density determines that bubbles cannot advance synchronously with water. Meanwhile, the disturbance from ambient flow structures alleviates the difference of bubble velocity in transverse direction.

5.5 Statistical Features of Bubble Volume

With the same sample region as that shown in Fig. 5.11, the probability density function (PDF) of bubble number is examined at various upstream velocities. The correspondence between the PDF of bubble number and bubble volume is shown in Fig. 5.11 and the air injection rate is 20 mL/min. The sample region covers two individual images that are continuous in streamwise direction. The result shown in Fig. 5.11 is obtained statistically based on two image groups, each contains 1024 images. The result is statistically stable.

The majority of bubbles is concentrated within a narrow volume range. As upstream velocity increases, the number of small bubbles increases apparently. In this context, bubble numbers near the upper and lower edges of the wake flow are expected to be multiplied since that small bubbles are abundant near wake edge.

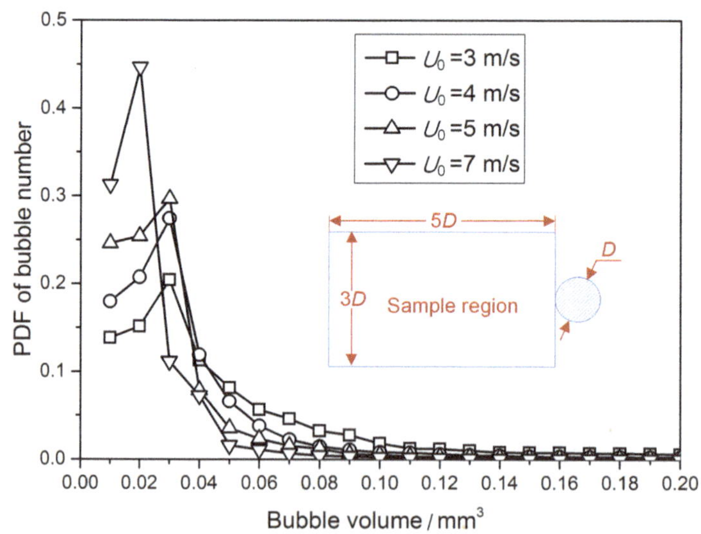

Fig. 5.11 Relationship between PDF of bubble number and bubble volume at air injection rate of 20 mL/min

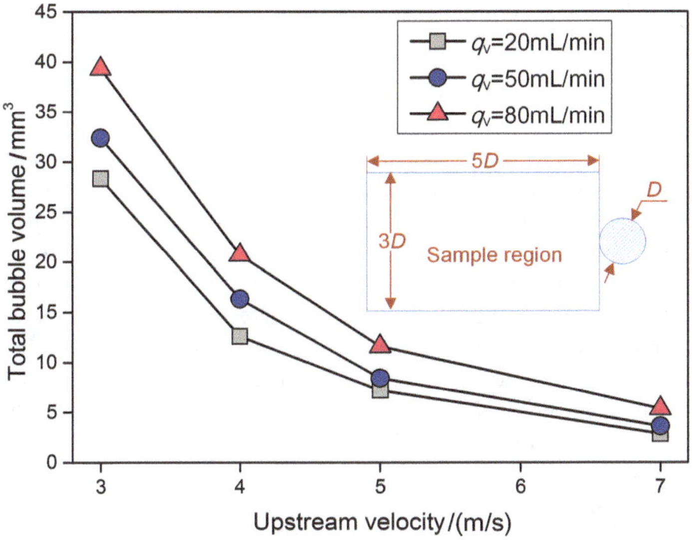

Fig. 5.12 The effect of air injection rate on bubble volume at 3 m/s

This viewpoint is in accordance with Fig. 5.9. Meanwhile, the variation of wake width with upstream velocity should be considered. Therefore, the bubbly flow pattern can sustain irrespective of the variation in the upstream velocity.

With the same sample region as that shown in Fig. 5.11, the total bubble volume is calculated for the four upstream velocities. In pipe flows, air void fraction could be readily calculated, according to the study documented in [24]. In Fig. 5.12, the influence of air injection rate is considered as well. As upstream velocity increases, total bubble volume decreases consistently, as is ascribed to the increase of the number of small bubbles and the collapse of large bubbles. Even as the air injection rate varies, such a tendency remains invariant. As air injection rate increases, total bubble volume rises, particularly at $U_0 = 3$ m/s. At high upstream velocities, the variation of total bubble volume with air injection rate is relatively slight. At high upstream velocity, bubbles generated near the injection hole are rapidly transported with the carrier flow and are then affected by vorticity and velocity gradients. It means that large bubbles cannot sustain for a long streamwise distance from the cylinder.

5.6 Concluding Remarks

In this chapter, bubbly flows downstream of a circular cylinder are produced by injecting air into the water tunnel through the flow passage inside the cylinder. An experimental study, employing particle image velocimetry and shadow image velocimetry, enables a comprehensive analysis of bubble parameters and their relationship with the carrier flow.

(1) Bubbly flow is in accordance with the carrier wake flow in terms of instantaneous meandering flow pattern. Relative to the carrier flow, bubbles have a small overall velocity magnitude and bubble velocity distribution in the middle subzone of the wake flow is uniform. Further downstream, the velocity deficit in the carrier flow diminishes, while bubble velocity distribution remains nearly invariant.
(2) Bubble size is inherently associated with vorticity distribution of the carrier flow, which shows dual-peak distribution pattern on the cross sections of the wake. Sauter mean diameter decreases drastically at the interface between the middle subzone and the upper or lower zone. The cross-sectional bubble size distribution is in reasonable accordance with the results obtained through the formula expressing the relationship between the turbulent kinetic energy dissipation rate and bubble size.
(3) As air injection rate remains constant, the percentage of small bubbles increases with upstream velocity. Small bubbles are dominant irrespective of the upstream velocity. As air injection rate increases, total bubble volume rises consistently. At high upstream velocity, total bubble volume is small and the augment of total bubble volume with air injection rate is relatively insignificant.

References

1. Brennen CE. Cavitation and bubble dynamics. Oxford: Oxford University Press; 1995.
2. Lee SJ, Kawakami E, Arndt REA. Measurements in the wake of a ventilated hydrofoil. In: ASME 2013 fluids engineering division summer meeting, Nevada, USA. 2013 July 7–11.
3. Xiang M, Cheung SPP, Yeoh GH, Zhang WH, Tu JY. On the numerical study of bubbly flow created by ventilated cavity in vertical pipe. Int J Multiph Flow. 2011;37:756–68.
4. Kawakami W, Arndt REA. Investigation of the behavior of ventilated supercavities. J Fluids Eng. 2011;133(9):091305.
5. Varaksin AY. Fluid dynamics and thermal physics of two-phase flows: problems and achievements. High Temp. 2013;51(3):377–407.
6. Akbar MHM, Hayashi K, Lucas D, Akio Tomiyama A. Effects of inlet condition on flow structure of bubbly flow in a rectangular column. Chem Eng Sci. 2013;104:166–176.
7. Colombet D, Legendre D, Cockx A, Guiraud P, Risso F, Daniel C, Galinat S. Experimental study of mass transfer in a dense bubble warm. Chem Eng Sci. 2011;66:3432–3440.
8. Hosokawa S, Tomiyama A. Bubble-induced pseudo turbulence in laminar pipe flows. Int J Heat Fluid Flow. 2013;40:97–105.
9. Karn A, Shao S, Arndt REA, Hong J. Bubble coalescence and breakup in turbulent bubbly wake of a ventilated hydrofoil. Exp Thermal Fluid Sci. 2016;70:397–407.
10. Batchelor GK. Pressure fluctuations in isotropic turbulence. Math Proc Cambridge Philos Soc. 1951;47:359–374.
11. Hinze JO. Fundamentals of the hydrodynamic mechanism of splitting in dispersion processes. AIChE J. 1955;1(3):289–295.

12. Rigby GD, Evans G, Jameson GJ. Bubble breakup from ventilated cavities in multiphase reactors. Chem Eng Sci. 1997;52(21–22):3677–3684.
13. Kelbaliev GI. Mass transfer between a drop or gas bubble and an isotropic turbulent flow. Theor Found Chem Eng. 2012;46(5):477–485.
14. Shawkat ME, Ching C, Shoukri M. On the liquid turbulence energy spectra in two-phase bubbly flow in a large diameter vertical pipe. Int J Multiph Flow. 2007;33(3):300–316.
15. Gore RA, Crowe CT. Effect of particle size on modulating turbulent intensity. Int J Multiph Flow. 1989;15(2):279–285.
16. Crowe CT. On models for turbulence modulation in fluid–particle flows. Int J Multiph Flow. 2000;26(5):719–727.
17. Hibiki T, Ishii M. Experimental study on interfacial area transport in bubbly two-phase flows. Int J Heat Mass Transf. 1999;42(16):3019–3035.
18. Michiyoshi I, Serizawa A. Turbulence in two-phase bubbly flow. Nucl Eng Des. 1986;95 (86):253–267.
19. Yeoh GH, Cheung SCP, Tu JY. On the prediction of the phase distribution of bubbly flow in a horizontal pipe. Chem Eng Res Des. 2012;90(1):40–51.
20. Liu TJ, Bankoff SG. Structure of air–water bubbly flow in a vertical pipe I. liquid mean velocity and turbulence measurements. Int J Heat Mass Transf. 1993;36(4):1049–1060.
21. Kataoka I, Serizawa A, Besnard DC. Prediction of turbulence suppression and turbulence modeling in bubbly two-phase flow. Nucl Eng Des. 1993;141(1–2):145–158.
22. Fujiwara A, Minato D, Hishida K. Effect of bubble diameter on modification of turbulence in an upward pipe flow. Int J Heat Fluid Flow. 2004;25(3):481–488.
23. Aloui F, Doubliez L, Legrand J, Souhar M. Bubbly flow in an axisymmetric sudden expansion: Pressure drop, void fraction, wall shear stress, bubble velocities and sizes. Exp Thermal Fluid Sci. 1999;19(2):118–130.
24. Sungkorn R, Derksen JJ, Khinast JG. Modeling of turbulent gas–liquid bubbly flows using stochastic Lagrangian model and lattice-Boltzmann scheme. Chem Eng Sci. 2011;66 (12):2745–2757.

Chapter 6
Drag-Type Hydraulic Rotor

Abstract The advantages of the drag-type rotor in wind energy utilization have been widely acknowledged. In recent years, the exploitation of water energy in off-shore regions, rivers or even pipes guides the application of the drag-type rotor into a new stage. In this chapter, a drag-type rotor operating in the medium of water is investigated. The water tunnel is used to furnish flow environment for the rotor. Flow patterns near the rotor are measured with particle image velocimetry technique and the wake flow is particularly emphasized. At various rotor setting angles and upstream velocity magnitudes, velocity and vorticity distributions in the wake flows are depicted and compared. The time-dependent torque coefficient of the rotor is calculated based on CFD results.

6.1 Introduction

The rotor can be used to absorb the kinetic energy of flowing fluid [1]. With water as the medium, the rotation of the drag-type rotor is exposed to considerable flow resistance [2]. Meanwhile, the medium inertia is high and will exert high hydraulic load on the rotor blades. Therefore, a favorable power output with the hydraulic drag-type rotor is expected.

Rotor structure is a primary factor underlying the operation capability of the hydraulic rotor [3]. The Darrieus rotor can also be used as a hydraulic rotor, and the lift exerted on the rotor blades with foil sections plays a critical role [4]. The deficiency of the lift-type rotor in this context is the low blade solidity which results in a portion of bulk flow passing by the rotor without sufficient energy transfer. The rotational speed of the rotor must be selected as fairly high. The drag-type rotor such as the Savonius rotor is immune from this issue. Additionally, the installation of the drag-type rotor is convenient and various environments can accommodate the application of the drag-type rotor [5]. The energy of flowing water in rivers or even pipes can be extracted using the drag-type rotor [6, 7]. Meanwhile, the size of the drag-type rotor in service covers a wide spectrum [8]. Theoretically, large rotor diameter is associated with a large amount of energy converted. Nevertheless, the

C. Kang et al., *Methods for Solving Complex Problems in Fluids Engineering*,
https://doi.org/10.1007/978-981-13-2649-3_6

enlarged rotor size entails high hydraulic load, so the material and structural strength of the rotor deserve particular attention [9]. In addition, increasing blade number of the drag-type rotor has also been attempted to harvest more water energy [10].

Flow characteristic and rotor performance is inherently correlated. To acquire flow parameter distributions near the drag-type rotor, both experimental and numerical approaches have been practiced. In terms of flow measurement using optical instruments such as particle image velocimetry, the rotor is required to be two dimensional, namely with identical cross sections along the rotor axis. As for rotors with twisted blades, local flows near the rotor cannot be detected completely [11]. Flow visualization is a qualitative experimental method, and complex flows near the rotor can be represented with tracers or streak lines [12]. Alternatively, the flow near of the drag-type rotor can be investigated with the CFD technique, which is prevalent in treating turbulent flows around the rotating rotor [13]. Furthermore, the pressure distribution over rotor blade surface obtained numerically is a prerequisite for calculating torque and power coefficients of the rotor.

The purpose of this chapter is to disclose flow characteristics of the drag-type hydraulic Savonius rotor. A drag-type rotor is devised based on the conventional Savonius rotor. Flow experiment is carried out in a water tunnel, and the rotor is mounted in the transparent test segment of the water tunnel. Velocity and vorticity distributions near the rotor are measured using particle image velocimetry technique. The effect of both rotor setting angle and upstream velocity on flow patterns is considered. Numerical simulation is performed to assess torque performance of the hydraulic rotor. The instantaneous torque coefficient is to be calculated as the rotor rotates. The correlation between flow and performance of the drag-type rotor is anticipated to be elucidated.

6.2 Experimental Setup

6.2.1 Drag-Type Hydraulic Rotor

The cross section of the hydraulic rotor investigated is displayed in Fig. 6.1. The rotor is a two-dimensional rotor with identical cross sections along the rotor axis. The rotor is similar with the Bach-type rotor [14]. Nevertheless, the two rotor blades are connected at the rotor hub to prevent the medium from flowing via the inter-blade gap. The rotor diameter is 56 mm and the tip-speed ratio (TSR) under the design condition is 0.3. In practice, the value of TSR often deviates from the designed value, so the optimum operation situation cannot be guaranteed [15]. It is fairly favorable provided that the rotor operates near the designed value of TSR. Furthermore, there is no available principle for guiding the determination of the TSR value with the medium of water.

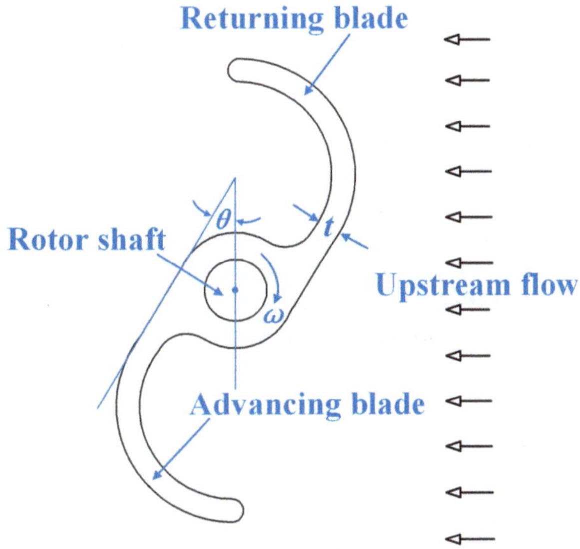

Fig. 6.1 Schematic view of the drag-type hydraulic rotor

In general, the blade thickness, *t*, is set identical over the Savonius rotor blade [16]. In the presented study, *t* keeps invariant from the rotor hub to blade tip. The rotor setting angle, θ, is defined as the relative angle between the vertical line passing through the center point of the rotor shaft and the tangent of the blade surface, as denoted in Fig. 6.1. Furthermore, as the rotor rotates, the interaction of upstream flow and the returning blade will hinder the clockwise rotation of the rotor. In some cases, the auxiliary device such as the deflector is used to alleviate the adverse effect [17].

The relationship between rotor performance and fluid flow is of significance for the drag-type rotor [18]. Regarding the rotor blade shape shown in Fig. 6.1, large-scale flow structures are anticipated downstream of the rotor. The geometrical characteristics of these flow structures depend to a large extent on the rotor orientation and upstream velocity. The visualization of these flow structures facilitates the improvement in rotor design [19]. Meanwhile, large-scale flow structures are associated with uneven flow parameter distributions, which affect rotor operation accordingly [20].

6.2.2 Test Segment of the Water Tunnel

Experimental facilities like the water tunnel, wave-generating flume, wind tunnel and tow-testing lorry have been utilized to test rotor performance [21–23]. Furthermore, the data obtained under experimental conditions furnish a sound support for the series design of rotors [24]. Another important experiment that can be carried out using these facilities is flow measurement. Nevertheless, relative to rotor performance test, flow measurement has rarely been reported [25]. Numerical simulation in this aspect is prevalent, particularly in recent years.

Here, the drag-type hydraulic rotor is installed in the transparent test segment of the water tunnel, and the rotor axis is vertical to the upstream flow direction. The test segment is identical with the one used in Chap. 5 and the four side walls of the test segment are transparent. In consideration of the rotor diameter and the width of the test segment, the blockage ratio is about 0.18. A stabilization segment situated upstream of the test segment is used to regulate upstream flow for the rotor and flow fluctuations will be suppressed as well [26]. The stabilization segment and the test segment is connected with a contraction segment, which is featured by a side wall profile conforming to the Witozinsky curve.

6.2.3 PIV Set-Up

Flow velocity distributions over the mid-span plane vertical to the rotor are measured using a LaVision PIV system, which is introduced in Chaps. 4 and 5. The configuration of the main components of the experimental rig is illustrated in Fig. 6.2. Hollow glass beads with diameters ranging from 10 to 20 μm are used as tracing particles. The quantity of the seeded tracing particles is gauged to a proper value, with which the detection of flow structures of both large and small scales in the wake flows considered is enabled; meantime, the tracing particles do not agglomerate at any local position. For the Darrieus rotor, low blade solidity facilitates the PIV measurement near individual blades [27]. By contrast, the rotor considered here is not transparent, so the flow regions on the upper and lower sides of the rotor cannot be illuminated simultaneously [28]. To deal with such a situation, two deployment schemes of the incident light lens are attempted, above and under the rotor. The latter scheme is shown in Fig. 6.2. The rotor is made of copper alloy and manufactured with the computerized numerical control (CNC) machining technique. Therefore, the agreement in rotor geometry between the practical rotor and the geometrical model is guaranteed [29].

During the experiment, the rotational speed of the circulating pump in the water tunnel is adjusted through a variable frequency controller and the upstream velocity, U_0, is varied between 2 and 7 m/s. Therefore, Reynolds number (*Re*) upstream of the rotor ranges from 1.2×10^5 to 4.1×10^5 with the characteristic length equivalent to the test segment width. At each upstream velocity, the rotor setting angle is set to 0°, 36°, 72°, 144° and 150°, respectively. The tracing particles are added into the water tank after the tunnel is deaerated. Furthermore, the transparent test segment is fully wetted by the medium and the influence of free surface wave is eliminated [30]. Regarding the experiment uncertainty, it is produced primarily in PIV measurement and is approximately 2.8%, which is calculated based on the specifications of the PIV system and the maximum velocity magnitude in the present experiment.

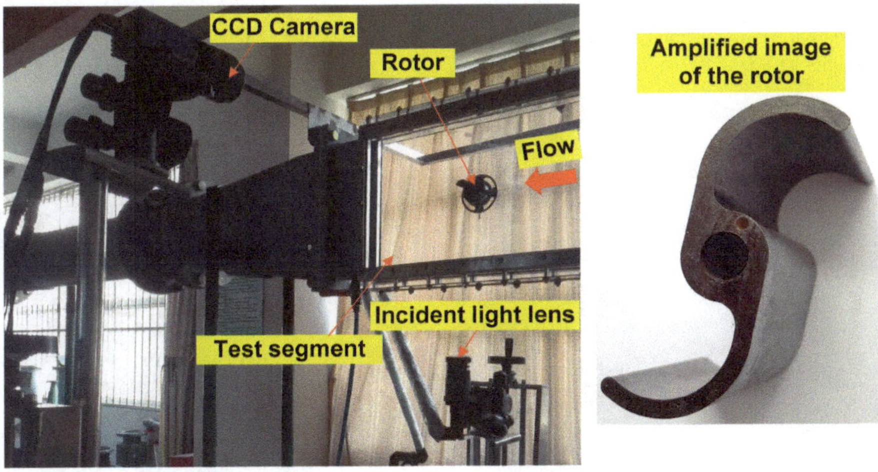

Fig. 6.2 Configuration of major components of the experiment system

6.3 Flow Patterns Near the Hydraulic Rotor

6.3.1 Wake Flow Patterns

Here, three upstream velocities, 2, 5 and 7 m/s are selected. At different rotor setting angles and the three upstream velocities, wake flow patterns downstream of the drag-type rotor are exhibited in Figs. 6.3, 6.4, 6.5 and 6.6. In this context, both velocity distribution and the streamlines are presented.

At $\theta = 0°$, with the variation of the upstream velocity, the rotor wake undergoes remarkable change. At U_0= 2 m/s, the maximum wake width even exceeds the rotor diameter, and in the wake region, three large vortices are distinct. Meanwhile, an isolated high-velocity region is situated in the middle part of the wake and immediately downstream of the rotor. As U_0 increases from 2 to 5 m/s, the wake pattern of triple vortices is replaced with the twin-vortex pattern. The wake is narrowed in transverse direction, as is clearly caused by the increase of the outer stream velocity. In terms of vortex size, the upper and lower vortices are nearly equivalent. The deviation of streamwise position between the two vortex cores is also obvious, as is reinforced at U_0= 7 m/s. In this connection, the upper vortex is propelled towards the wake centerline and the elongated vortex profile is transformed into a seemingly circular shape. In general, violent turbulent fluctuations occur at the upper and lower edges of the wake region [31]. In Fig. 6.3, high velocity gradients at the interface between the rotor wake and the outer stream contribute significantly to the occurrence of turbulent shear effect and then turbulent fluctuations.

The wake flow patterns corresponding to $\theta = 36°$ are shown in Fig. 6.4. In terms of velocity distribution at the interface between outer stream and the wake region, low upstream velocity is responsible for low velocity gradients. In Fig. 6.4c, the demarcation between the wake region and outer stream is apparent, and the

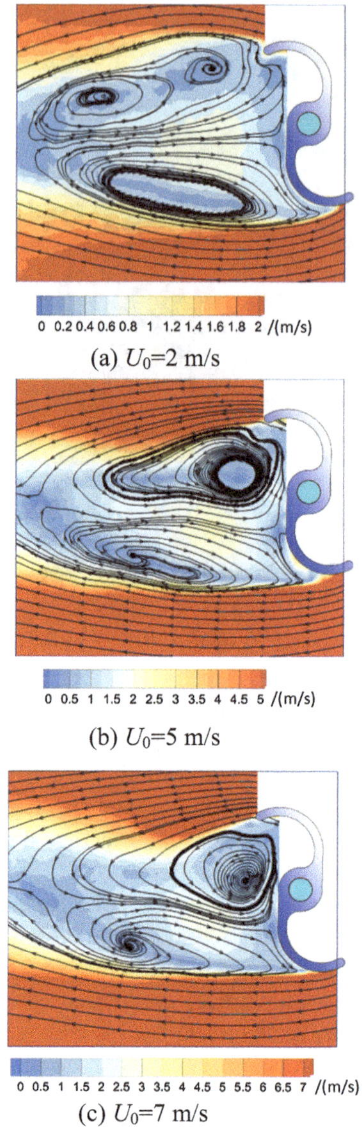

Fig. 6.3 Wake flow patterns at different upstream velocities at $\theta = 0°$

streamwise extension of the wake is enhanced relative to Fig. 6.4a, b. As for the development of large-scale flow structures in the wake region, the evolution of the upper vortex is similar with that illustrated in Fig. 6.3, the vortex core approaches wake center as the upstream velocity increases. It is noticeable in Fig. 6.4b that the wake region is dominated by a single vortex, while the other subfigures are featured by dual-vortex pattern. In Fig. 6.4c, the interaction between the two large vortices is intensified and the two vortex boundaries are jointed together near the convex side of the advancing blade.

Fig. 6.4 Wake flow patterns at different upstream velocities at $\theta = 36°$

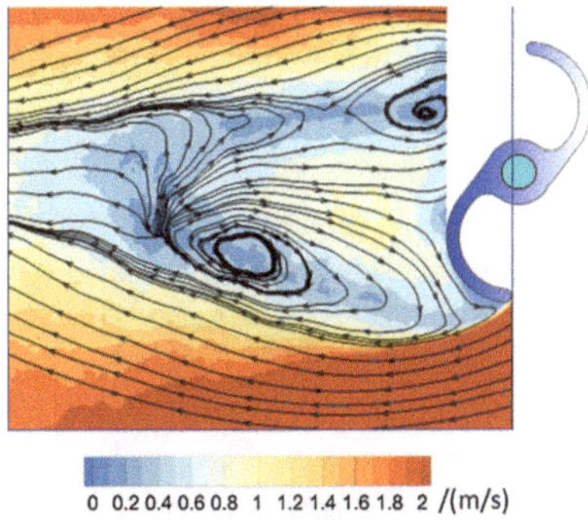

(a) U_0=2 m/s

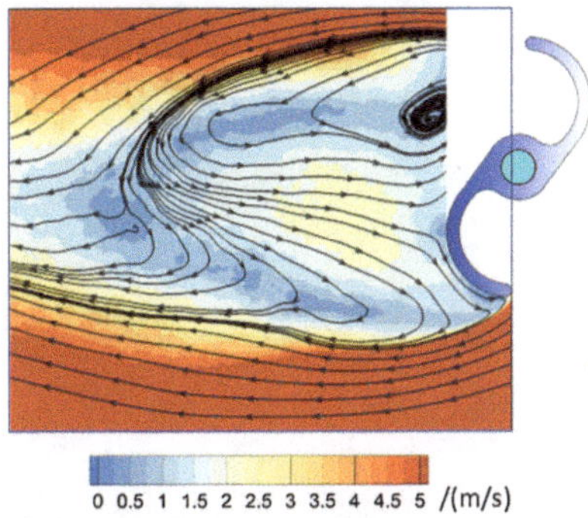

(b) U_0=5 m/s

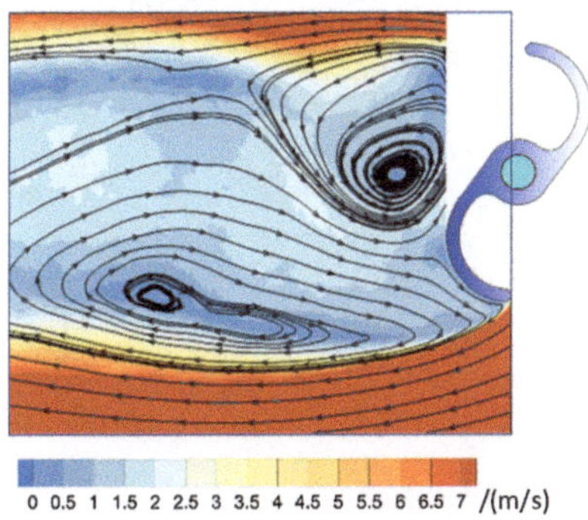

(c) U_0=7 m/s

The influence of rotor blades on the wake vortex profile is prominent. Particularly, near the rotor, the streamline shape conforms explicitly to the blade surface. In turn, the flow near the rotor blades will exert a hydraulic load on the blade surface, which is directly related to the torque or power output of the rotor.

The three wake patterns at the rotor setting angle of 72° are displayed in Fig. 6.5. Relative to Figs. 6.3 and 6.4, the width of the wake is reduced obviously in Fig. 6.5, as is inseparable with the orientation of the rotor. Overall, the three subfigures included in Fig. 6.5 are rather analogous. Not just the shape of the two

Fig. 6.5 Wake flow patterns at different upstream velocities at $\theta = 72°$

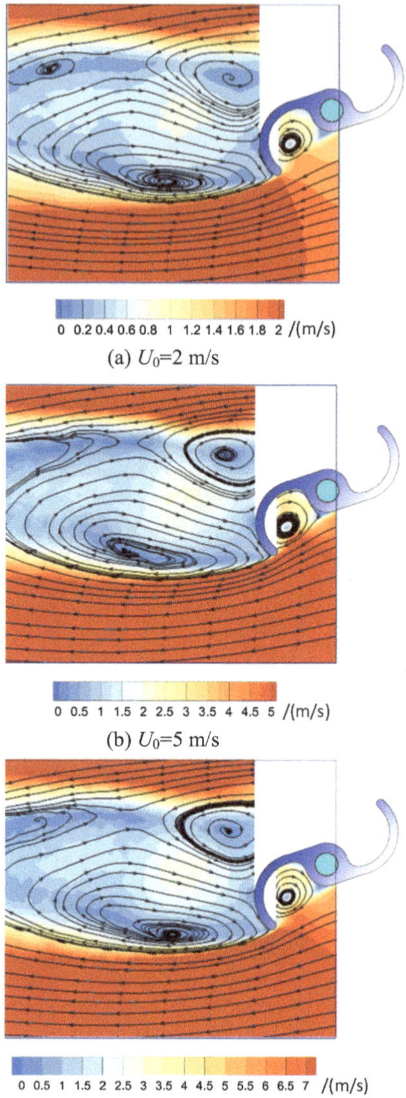

(a) U_0=2 m/s

(b) U_0=5 m/s

(c) U_0=7 m/s

large vortices but also the position of the vortex core show only a slight variation as the upstream velocity increases. In addition, there is a vortex trapped by the concave side of the advancing blade, and such a vortex almost covers the entire semi-circular region. Moreover, the vortex remains stable with the increase of upstream velocity. In this context, the unique solid boundary provided by the rotor blade, in conjunction with the outer stream induces such a vortex. As the outer stream velocity increases, the shear effect between the outer stream and the stagnant medium in the semi-circular region is reinforced, and the vorticity is improved

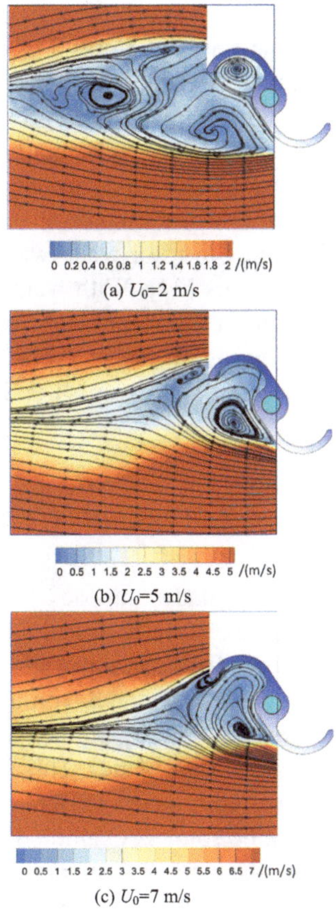

Fig. 6.6 Wake flow patterns at different upstream velocities at θ = 144°

accordingly. Nevertheless, the spatial expansion of the vortex is restricted by the blade. From another perspective, energy losses associated with the flow structures shown in Fig. 6.5 are considerable.

As the rotor setting angle increases to 144°, the flow patterns at the three upstream velocities are displayed in Fig. 6.6. At U_0= 2 m/s, the wake width is fairly large and in streamwise direction, two large vortices are distinct. Another vortex is confined in the semi-circular region under the returning blade, as is similar to that shown in Fig. 6.5. As the upstream velocity increases, the wake attenuates and the low-velocity area is apparently narrowed, as shown in Fig. 6.6b, c. In the wake regions shown in the two figures, large-scale vortices are absent. The only vortex identifiable in Fig. 6.6b, c is situated near the convex side of the advancing blade, the intrusion of the vortex into the semi-circular area associated with the returning blade results in the collapse of the confined vortex displayed in Fig. 6.6a. In Fig. 6.6, the wake flow region is connected with the local flow region under the returning blade. As the upstream velocity increases, the curvature of the streamlines adjacent to the convex side of the advancing blade changes apparently, leading to contracted wake flow region and intensified interaction between the rotor blades and the flow structures.

Flow patterns serve as an unambiguous base for the optimization of the drag-type rotor geometry [32]. Here, the rotor is two-dimensional, so the planar flow pattern is representative. As for the drag-type rotor equipped with twisted blades, the flow patterns are spatially uneven and spinning vortices develop in both streamwise and transverse directions [33]. In addition, as the rotor rotates, the flow structures in rotor wake will alter periodically, as has been proved in the study of wind rotors [34]. With water as the medium, the response of flow structures to the rotation of the rotor might be rather delayed in consideration of the large medium density. Finally, the wake attenuates and the influence of flow structures to ambient environment will vanish entirely.

6.3.2 Flow Characteristics Near the Rotor

The rotor blades impose a boundary effect on the upstream flow [35]. Such an effect might be alleviated or reinforced as the rotor rotates. The representative of complex upstream flow encountered by the drag-type rotor is the wake-type upstream flow, which occurs as the Savonius rotor is surrounded by the Darrieus rotor blades [36]. In this case, the originally complex upstream flow interacts with the rotating Savonius rotor blades. This situation cannot be explained with common knowledge of the drag-type rotor. Here, the influence of upstream velocity on the flows near the hydraulic rotor is examined.

At the rotor setting angle of 150°, the incident light lens is deployed above and under the rotor, respectively. Then flows surrounding the rotor was measured at different upstream velocities, and the result is shown in Fig. 6.7, where flow direction is from the right to the left. In Fig. 6.7a, b, flow velocity distributions are rather similar in terms of both upstream flow and the local flows adjacent to the rotor blades. In the semi-circular regions enclosed by the concave sides of the rotor blades, small-scale flow structures are explicit. With respect to these regions, the flow velocity near the advancing blade is obviously high compared to that associated with the returning blade. In Fig. 6.7c, the upstream flow pattern and the flow trapped by the concave side of the advancing blade are analogous with those shown in Fig. 6.7a, b. The uniqueness of Fig. 6.7c lies in the outer stream beneath the advancing blade, it travels along the convex surface and then shifts towards the top-left after a sharp deflection. As a consequence, the wake is considerably narrowed relative to the other two subfigures, and meanwhile, vortices are fostered in the semi-circular region surrounded by the returning blade, as can be extrapolated from Fig. 6.7c.

Similar to the drag-type wind rotor, the influence of blade shape and overlap ratio on flow and rotor performance is non-negligible [37]. Provided that the gap between the two blades is not filled, a portion of fluid will pass through the gap and influence the flow at the other side [38]. Near-rotor flow, as well as rotor performance will undergo a remarkable change.

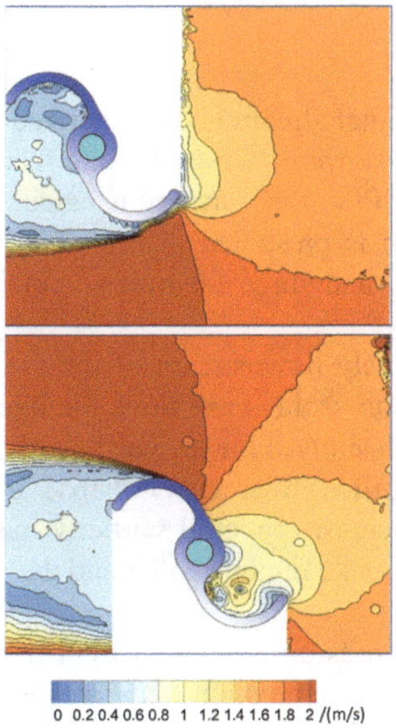

(a) U_0=2 m/s

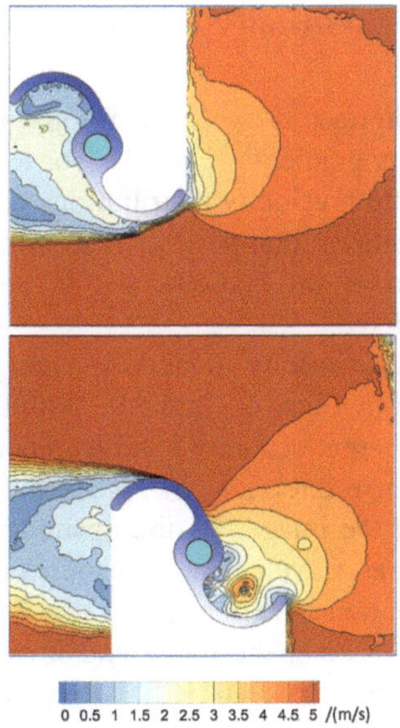

(b) U_0=5 m/s

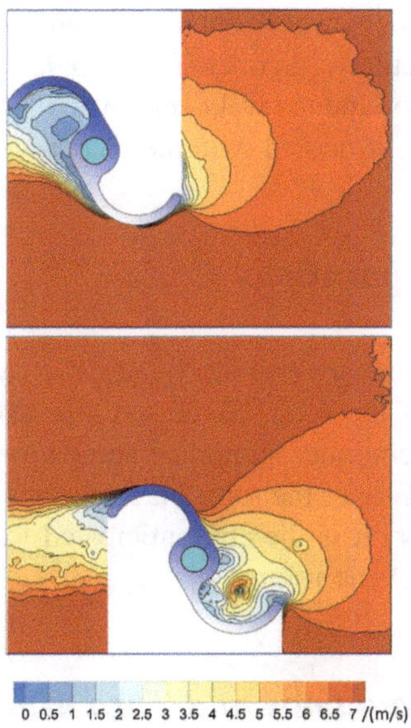

(c) U_0=7 m/s

Fig. 6.7 Velocity distributions near the rotor at different upstream velocities (θ = 150°)

6.3.3 Vorticity Distribution

The interface between the rotor wake and the outer stream is characterized by high velocity gradients. Thus the vorticity might be concentrated at the interface. Based upon flow velocity distributions acquired with PIV, the vorticity is calculated and the vorticity component in rotor axis direction is presented in Figs. 6.8 and 6.9. Two rotor orientations, $\theta = 0°$ and $\theta = 150°$, and three upstream velocities are considered.

High vorticity magnitude emerges immediately downstream of the rotor, as is shared by Figs. 6.8 and 6.9. With the progression of the wake flow, the integrity of the coherent vorticity band is undermined consistently. Finally, vorticity diffuses in transverse direction. In Figs. 6.8a and 6.9a, vorticity elements collapse at the left end of the image. As the upstream velocity increases, overall vorticity magnitude increases as well. The correspondence between wake region profile and the vorticity distribution is explicit.

In Fig. 6.8, the upper and lower vorticity bands are separated by a fairly wide region with low vorticity, which signifies that the interaction between the vortices fostered in upper and lower vorticity bands is mitigated. Consequently, small flow structures generated under this mechanism are absent near wake centerline. By contrast, in Fig. 6.9b, c, due to the narrowed wake, the two vorticity bands influence each other apparently, then the wake center is dominated by small vorticity elements. Such a situation is particularly distinct in Fig. 6.9c. In this connection, the rotor severs as a vortex generator and the wake region is filled with shedding vortices. The propagation of vortices is promoted with the increase of the upstream velocity.

6.4 Numerical Preparations

Static pressure distributions over rotor blade surface can be used to estimate operation performance of the rotor. Since static pressure cannot be acquired with the flow measurement technique, numerical simulation is performed instead. Then static pressure distribution over the rotor blade surface is integrated to calculate the torque output of the rotor. It is thereby anticipated to correlate rotor performance and flow characteristics of the rotor.

6.4.1 Governing Equations and Turbulence Model

In parallel with the experimental condition, pure water of 20 °C is used as the flow medium in the simulation. In consideration of the side wall effect, the fluid flow is deemed as three-dimensional in spite of the two-dimensional rotor. Therefore, the incompressible flow near the hydraulic rotor is governed by Reynolds averaged Navier-Stokes (RANS) equations, including continuity and momentum equations [39].

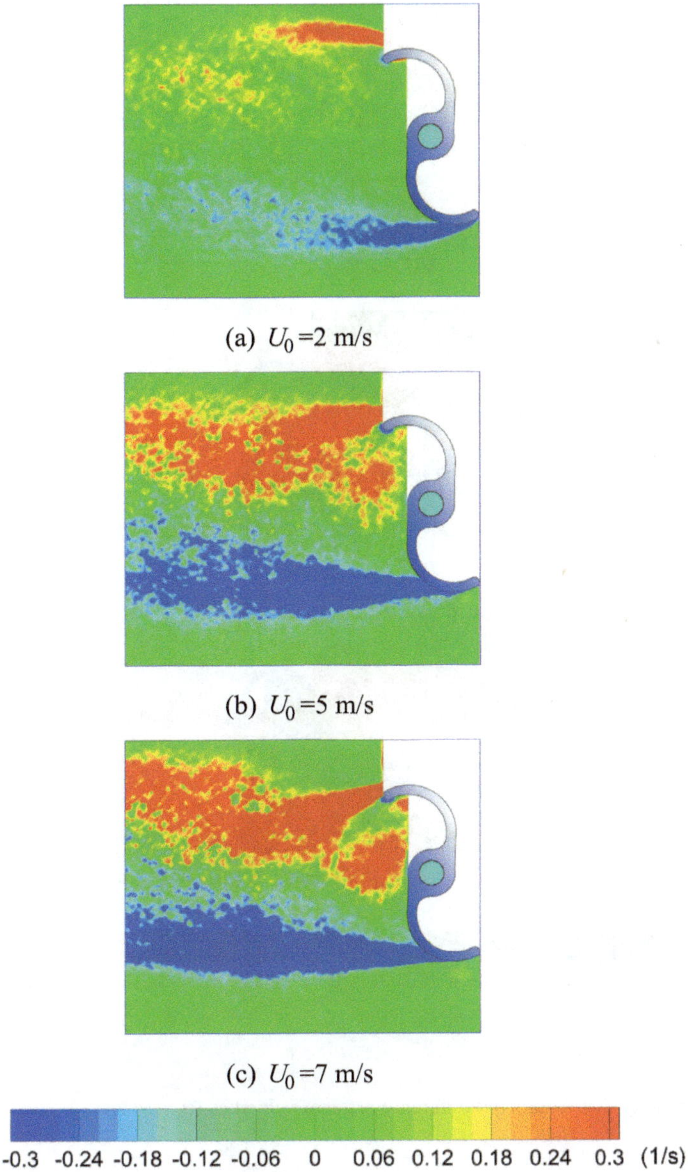

Fig. 6.8 Vorticity distributions at different upstream velocities (θ = 0°)

Here, the shear stress transport (SST) k–ω turbulent model is selected. In [40], four turbulent models, re-normalization group (RNG) k–ε, realizable k–ε, SST k–ω and standard k–ε models are compared with flows near the Savonius rotor. The SST

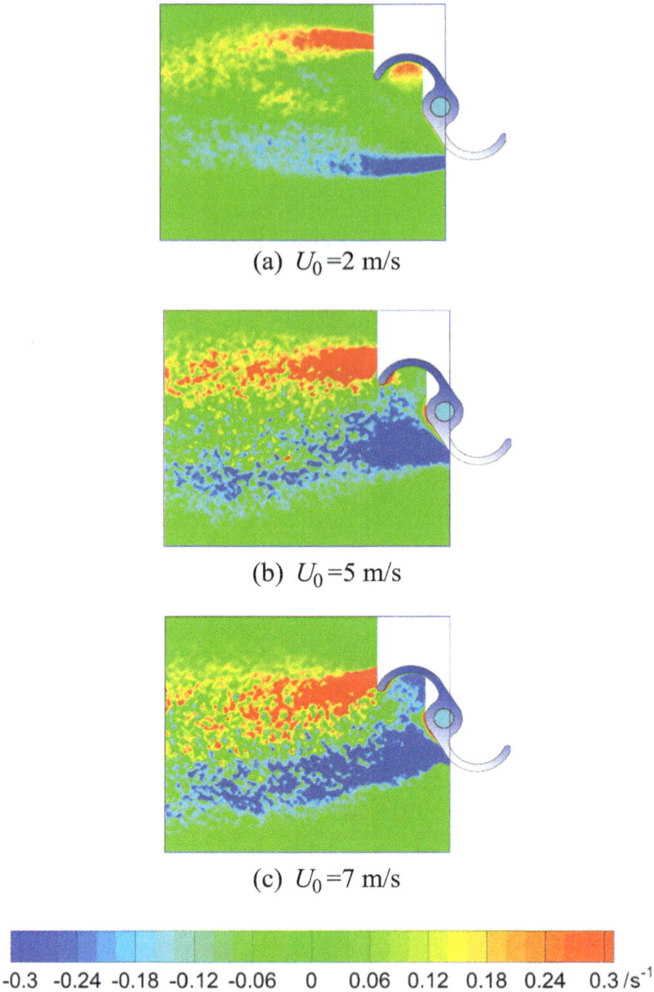

(a) U_0=2 m/s

(b) U_0=5 m/s

(c) U_0=7 m/s

Fig. 6.9 Vorticity distributions at different upstream velocities (θ = 144°)

k–ω turbulent model proves to be the most feasible one in this case. Moreover, the SST k–ω turbulence model combines the advantages of k–ε and k–ω models and can properly treat both the outer flow and the near-wall flows relative to the rotor blade.

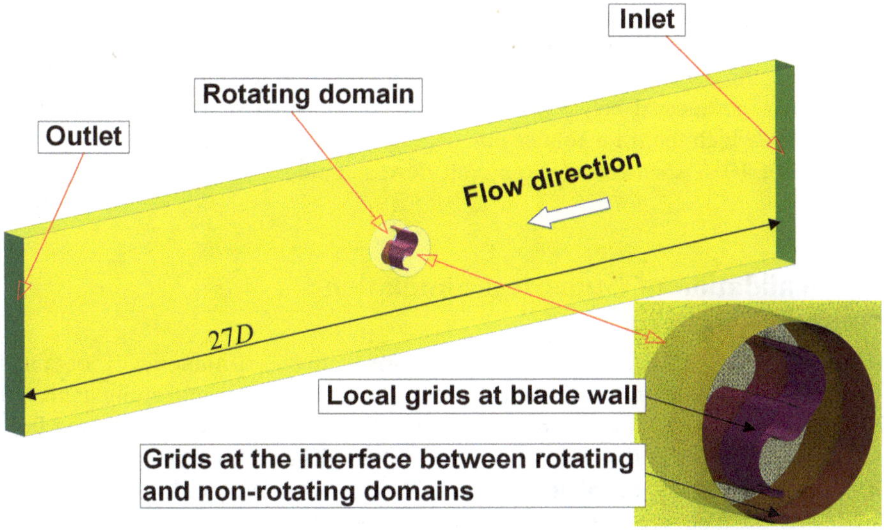

Fig. 6.10 Schematic diagram of the computational domain

6.4.2 Grid Deployment Scheme

The three-dimensional computational domain with dimensions of 1500 mm 315 mm × 50 mm is constructed and displayed in Fig. 6.10. The rotor rotation subdomain bounded by the rotor blade surface is included in the domain. The rotor geometry is identical with that of the practical rotor. The influence of the side wall of the test segment on fluid flow is critical. Therefore, the agreement between numerical and experimental schemes in this aspect is guaranteed [41]. The grids near the rotor blade surface are refined to identify near-wall flow phenomena [42]. The whole computational domain is discretized with about 3.8 million unstructured tetrahedral grids, and local grids are illustrated in Fig. 6.10 as well. The equiangle skewness is 0.22. The values of y^+ range from 27 to 61. For the interface between rotating and non-rotating domains, the fidelity of data exchange is ensured through adjusting grid density near the interface.

6.4.3 Boundary Conditions

The commercial code ANSYS CFX is employed as the numerical solver. No-slip boundary condition is set at the rotor wall and the side wall of the test segment. The surface roughness of the rotor blade and the side wall is set to 0.025 and 0.05 mm, respectively. Near-wall flow regions are treated with scalable wall functions. Velocity inlet boundary condition is defined for the inlet of the computational domain, and pressure outlet boundary condition for the outlet. The rotational speed of the rotor is set according to the value of TSR. The sliding mesh method is used to accomplish the synchronous rotation of the grids in the rotating domain with the rotor. The time interval between neighboring computation steps is set to the time span within which the rotor rotates one degree. At each time step, the convergence threshold of 10^{-4} is set for all monitored flow quantities.

6.5 Validation of Numerical Simulation

To validate the numerical strategy, a comparison between numerically obtained flow patterns and flow measurement result is implemented at rotor setting angles of 72° and 150° for the same upstream velocity of 5 m/s. Numerically obtained flow velocity distribution at $\theta = 72°$ is displayed in Fig. 6.11a. Experimentally obtained upstream flow velocity distribution at $\theta = 72°$ is shown in Fig. 6.11b. In Fig. 6.11a, near the rotor, there are five characteristic zones. As for Zone A, situated upstream of the rotor, it is similar with that shown in Fig. 6.11b in terms of both velocity distribution pattern and velocity magnitude. Zone B involves a low-velocity band which is contiguous to Zone C. Among all the five zones, Zone C has moderately high velocity and is near the wake center. Zone D covers a tilted velocity band with layers of different velocity magnitudes. The configuration of Zones B, C and D is also identifiable in Fig. 6.5a. Meanwhile, Zone B and Zone E is connected in a seemingly intermittent fashion, as is shared by Figs. 6.5a and 6.11a. The comparison between numerically and experimentally obtained large-scale flow structures demonstrates that the numerical scheme is reliable.

The similarity between Fig. 6.12a, b is explicit in view of the vortices trapped by the advancing and returning blades. With the numerical scheme adopted, the joint action of the outer stream and rotor blades is predicted accurately. The deviation between numerical and experimental results is the wall effect associated with rotor blades. More specifically, as flow develops near the rotor blades, the numerically and experimentally obtained positions where the flow separates from the blade surface are different. This triggers the different in vortex size and vortex core position, as can be seen near the concave side of the returning blade and near the concave side of the advancing blade, respectively.

In general, the power output and the energy conversion ratio of the drag-type rotor are related mainly with large-scale flow structures [43]. Thus a reasonable numerical detection of large-scale flow structures ensures the agreement of performance coefficient between numerical and experimental results [44]. In the light of the essence of the turbulent flow, the interaction among wake vortices downstream of the rotor with multiple scales is intense [45]. Such an interaction cannot be adequately depicted in numerical results. Numerical result is influenced by multiple factors such as numerical algorithms, turbulent model, grids and surface roughness. Hitherto, it is still impractical to scale independently and quantitatively the contribution of any individual factor.

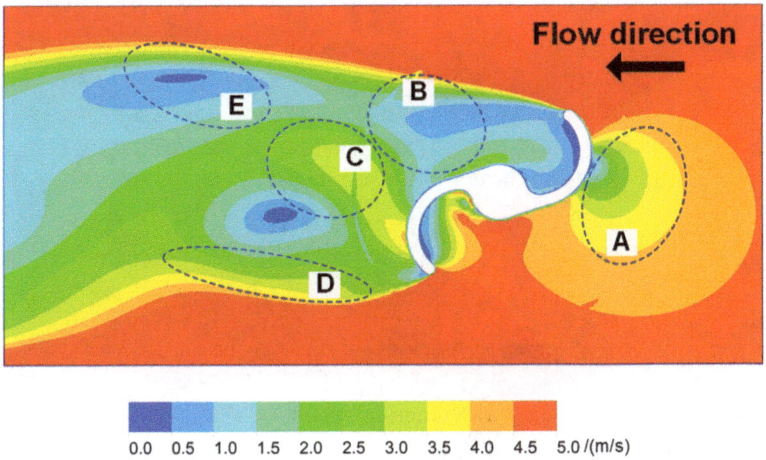

(a) Flow pattern obtained numerically

(b) Upstream flow velocity distribution measured

Fig. 6.11 Flow velocity comparison at $\theta = 72°$

6.6 Variation of Torque Coefficient with Rotor Rotation

Based on numerically obtained static pressure distribution over the rotor blade surface, the torque generated by the hydraulic load exerted on the rotor can be calculated. Unsteady simulation yields time-dependent pressure distributions, which are used to acquire instantaneous torque coefficient. At the three upstream flow velocities considered and the same value of TSR, the variation of torque coefficient in one full rotor rotation cycle is plotted in Fig. 6.13.

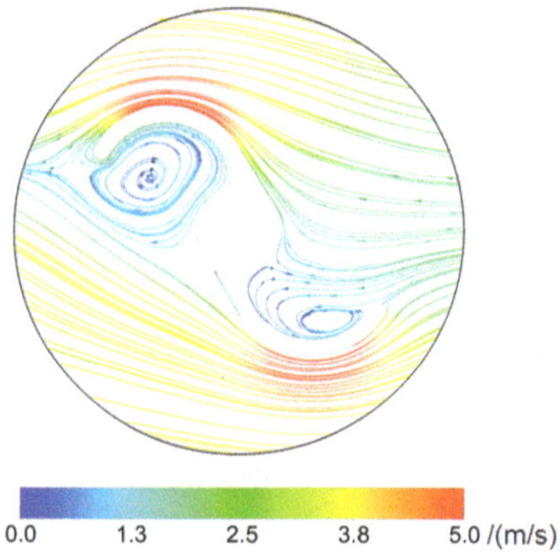

(a) Numerical result

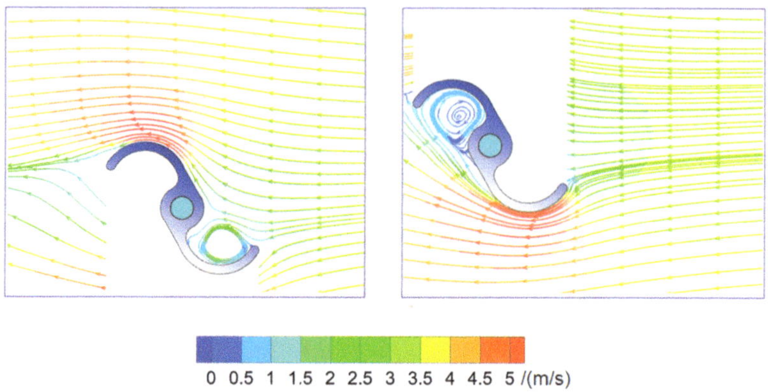

(b) Experimental result

Fig. 6.12 Flow structures near the rotor at θ = 150°

Within a full rotor rotation cycle, high torque coefficients are concentrated in the first and the third quadrants, as is shared by the three subfigures. At U_0= 5 and 7 m/s, the torque coefficient distribution tendencies are nearly identical. The cashew-type distribution patterns shown in Fig. 6.13 are similar to that reported in [46]. It is noticeable that in two ranges of rotor setting angle, one is near 120° and the other one is near 300°, the torque coefficient is low. At U_0= 5 and 7 m/s, torque coefficient fluctuation is apparent. In this context, the density of water is large, so the response of water to the variation of rotor setting angle is fairly delayed. Near the two rotor setting angles, the rotor is in an adverse state in terms of clockwise rotation, as can be inferred from the orientation of the rotor. Away from the two ranges of rotor setting angle, torque coefficient increases and shrinks in a regular manner. It is reported in [47] that at low upstream velocity less than 1 m/s, the fluctuation of torque within the entire rotation cycle is limited. Here, at U_0= 2 m/s,

Fig. 6.13 Time-dependent torque coefficient at different upstream velocities

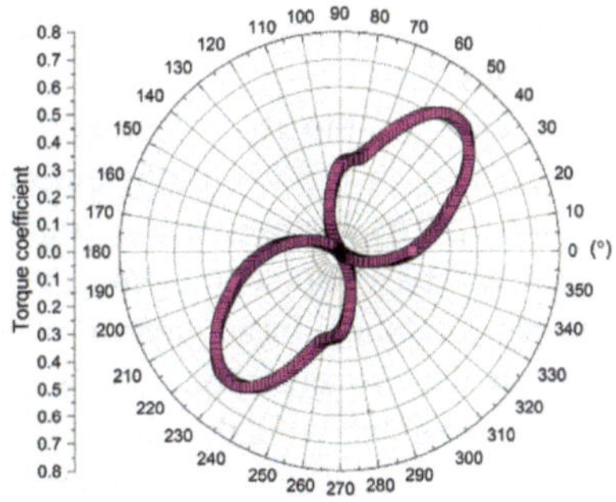

(a) U_0=2 m/s

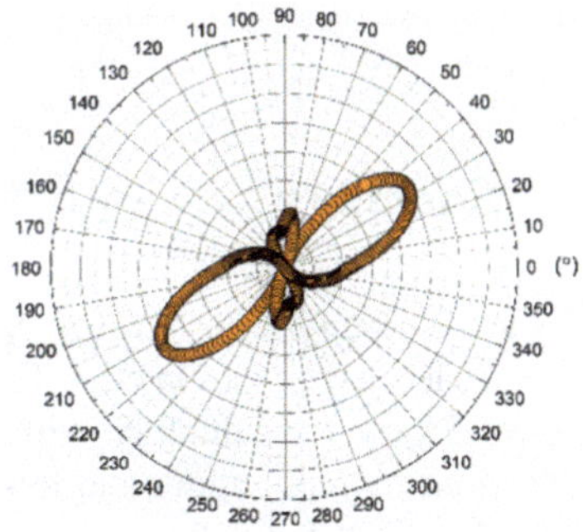

(b) U_0=5 m/s

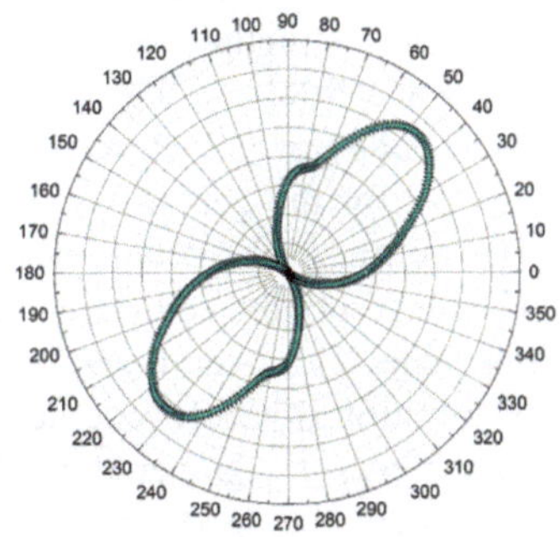

(c) U_0=7 m/s

the variation of the torque coefficient is relatively smooth relative to the other situations.

In view of the rotor studied, as it operates in the medium of air, the rotor possesses high torque coefficient compared with the conventional drag-type rotor does [48]. The overall torque coefficient shown in Fig. 6.13 is high and the maximum torque coefficient attains 0.7. This is inseparable from the large medium density and then large inertia which is transformed into the impetus for rotor rotation. The largest torque coefficient appears near the rotor setting angle of 30°, as is in agreement with the flow situation demonstrated in Fig. 6.4. Therefore, the correlation between flow pattern and torque performance is apparent. Similar rationale applies to the situation at the rotor setting angle of 150°, at which the rotor rotation is subjected to a balance between impetus and resistance, and the rotation inertia of the rotor turns to be the main factor that sustains the rotation of the rotor.

6.7 Summary

(1) Flow velocity distribution near a drag-type hydraulic rotor is obtained using particle image velocimetry. The width and streamwise development of the rotor wake vary considerably with the rotor setting angle. Inside the wake, three regions, namely upper and lower regions with low velocity, middle region with fairly high velocity, are typical. Immediately upstream of the rotor, the upstream flow attenuates rapidly due to the resistance of the rotor blades.

(2) Apart from the large vortices trapped by the concave sides of the rotor blades, large flow structures are produced evidently by the interaction of outer stream and wake flow. As the wake region is narrowed, the communication between large-scale flow structures is reinforced. Vorticity distribution furnishes a sound support for the generation mechanism of small vortices near the wake centerline. As the wake progresses, vorticity elements collapse finally.

(3) The correspondence between flow pattern and torque coefficient is explicit and traceable. The overall torque coefficient of the drag-type rotor is high, but the distribution of torque coefficient over the rotation cycle is uneven. High torque coefficients are gathered in the first and third quadrants, as is retained irrespective of the upstream velocity. The feature of torque coefficient is associated with large medium density and delayed medium response to rotor rotation.

References

1. Laws ND, Epps BP. Hydrokinetic energy conversion: technology, research, and outlook. Renew Sustain Energy Rev. 2016;57:1245–1259.
2. Lago LI, Ponta FL, Chen L. Advances and trends in hydrokinetic turbine systems. Energy Sustain Dev. 2010;14:287–296.

3. Vermaak HJ, Kusakana K, Koko SP. Status of micro-hydrokinetic river technology in rural applications: a review of literature. Renew Sustain Energy Rev. 2014;29:625–633.
4. Kirke BK. Tests on ducted and bare helical and straight blade Darrieus hydrokinetic turbines. Renew Energy. 2011;36(11):3013–3022.
5. Al-Bahadly I. Building a wind turbine for rural home. Energy Sustain Dev. 2009;13:159–165.
6. Sahim K, Santoso D, Radentan A. Performance of combined water turbine with semielliptic section of the Savonius rotor. Int J Rotating Mach. 2013;2013, Article ID: 985943.
7. Chen J, Yang HX, Liu CP, Lau CH, Lo M. A novel vertical axis water turbine for power generation from water pipelines. Energy. 2013;54:184–193.
8. Ikeda T, Iio S, Tatsuno K. Performance of nano-hydraulic turbine utilizing waterfalls. Renew Energy. 2010;35(1):293–300.
9. Kumar D, Sarkar S. Numerical investigation of hydraulic load and stress induced in Savonius hydrokinetic turbine with the effects of augmentation techniques through fluid-structure interaction analysis. Energy. 2016;116:609–618.
10. Thiyagaraj J, Rahamathullah I, SureshPrabu P. Experimental Investigations on the performance characteristics of a modified four bladed Savonius hydro-kinetic turbine. Int J Renew Energy Res. 2016;6(4):1530–1536.
11. Kamoji MA, Kedare SB, Prabhu SV. Performance tests on helical Savonius rotors. Renew Energy. 2009;34(3):521–529.
12. Miyoshi N, Shouichiro I, Toshihiko I. Performance of Savonius rotor for environmentally friendly hydraulic turbine. J Fluid Sci Technol. 2008;3(3):420–429.
13. McTavish S, Feszty D, Sankar T. Steady and rotating computational fluid dynamics simulations of a novel vertical axis wind turbine for small-scale power generation. Renew Energy. 2012;41:171–179.
14. Zhou T, Rempfer D. Numerical study of detailed flow field and performance of Savonius wind turbines. Renew Energy. 2013;51:373–378.
15. Rosmin N, Jauhari AS, Mustaamal AH, Husin F, Hassan MY. Experimental study for the single-stage and double-stage two-bladed Savonius micro-sized turbine for rain water harvesting (RWH) system. Energy Procedia. 2015;68:274–281.
16. Damak A, Driss Z, Abid MS. Experimental investigation of helical Savonius rotor with a twist of 180°. Renew Energy. 2013;52:136–142.
17. Iio S, Katayama Y, Uchiyama F, Sato E, Ikeda T. Influence of setting condition on characteristics of Savonius hydraulic turbine with a shield plate. J Thermal Sci. 2011;20(3):224–228.
18. Akwa JV, Vielmo HA, Petry AP. A review on the performance of Savonius wind turbines. Renew Sustain Energy Rev. 2012;16(5):3054–3064.
19. Roy S, Saha UK. Wind tunnel experiments of a newly developed two-bladed Savonius-style wind turbine. Appl Energy. 2015;137:117–125.
20. Kumar A, Saini RP. Performance parameters of Savonius type hydrokinetic turbine—a review. Renew Sustain Energy Rev. 2016;64:289–310.
21. Harries T, Kwan A, Brammer J, Falconer R. Physical testing of performance characteristics of a novel drag-driven vertical axis tidal stream turbine; with comparisons to a conventional Savonius. Int J Mar Energy. 2016;14:215–228.
22. Goh SC, Boopathy SR, Krishnaswami C, Schlüter JU. Tow testing of Savonius wind turbine above a bluff body complemented by CFD simulation. Renew Energy. 2016;87:332–345.
23. Rafiuddin Ahmed M, Faizal M, Prasad K, Cho Y-J, Kim C-G, Lee Y-H. Exploiting the orbital motion of water particles for energy extraction from waves. J Mech Sci Technol. 2010; 24(4):943–949.
24. Khan MNI, Iqbal T, Hinchey M, Masek V. Performance of Savonius rotor as a water current turbine. J Ocean Technol. 2009;4(2):71–83.
25. Danao LA, Eboibi O, Howell R. An experimental investigation into the influence of unsteady wind on the performance of a vertical axis wind turbine. Appl Energy. 2013;107:403–411.
26. Kailash G, Eldho TI, Prabhu SV. Performance study of modified Savonius water turbine with two deflector plates. Int J Rotating Mach 2012;2012, Article ID 679247.

27. Edwards JM, Danao LA, Howell RJ. PIV measurements and CFD simulation of the performance and flow physics and of a small-scale vertical axis wind turbine. Wind Energy. 2015;18(2):201–217.
28. Fujisawa N, Gotoh F. Visualization study of the flow in and around a Savonius rotor. Exp Fluids. 1992;12(6):407–412.
29. Altan BD, Altan G, Kovan V. Investigation of 3D printed Savonius rotor performance. Renew Energy. 2016;99:584–591.
30. Rafiuddin Ahmed M, Faizal M, Lee Y-H. Optimization of blade curvature and inter-rotor spacing of Savonius rotors for maximum wave energy extraction. Ocean Eng. 2013;65:32–38.
31. Akinari S, Yuichi M, Yuji T, Yasushi T. Interactive flow field around two Savonius turbines. Renew Energy. 2011;36(2):536–545.
32. Kang C, Liu H, Xin Y. Review of fluid dynamics aspects of Savonius-rotor-based vertical-axis wind rotors. Renew Sustain Energy Rev. 2014;33:499–508.
33. Lee J-H, Lee Y-T, Lim H-C. Effect of twist angle on the performance of Savonius wind turbine. Renew Energy. 2016;89:231–244.
34. Wang L, Yeung RW. On the performance of a micro-scale Bach-type turbine as predicted by discrete-vortex simulations. Appl Energy. 2016;183:823–836.
35. Faizal M, Rafiuddin Ahmed M, Lee Y-H. On utilizing the orbital motion in water waves to drive a Savonius rotor. Renew. Energy. 2010;35(1):164–169.
36. Ghosh A, Biswas A, Sharma KK, Gupta R. Computational analysis of flow physics of a combined three bladed Darrieus Savonius wind rotor. J Energy Inst. 2015;88(4):425–437.
37. Al-Kayiem HH, Bhayo BA, Assadi M. Comparative critique on the design parameters and their effect on the performance of S-rotors. Renew Energy. 2016;99:1306–1317.
38. Nasef MH, El-Askary WA, AbdEL-hamid AA, Gad HE. Evaluation of Savonius rotor performance: static and dynamic studies. J Wind Eng Ind Aerodyn. 2013;123:1–11.
39. Sharma S, Sharma RK. Performance improvement of Savonius rotor using multiple quarter blades—a CFD investigation. Energy Convers Manage. 2016;127:43–54.
40. Nasef MH, El-Askary WA, AbdEL-hamid AA, Gad HE. Evaluation of Savonius rotor performance: static and dynamic studies. J Wind Eng Ind Aerodyn. 2013;123:1–11.
41. Frikha S, Driss Z, Ayadi E, Masmoudi Z, Abid MS. Numerical and experimental characterization of multi-stage Savonius rotors. Energy. 2016;114:382–404.
42. Lam HF, Peng HY. Study of wake characteristics of a vertical axis wind turbine by two-and three-dimensional computational fluid dynamics simulations. Renew Energy. 2016;90:386–398.
43. Fujisawa N. Velocity measurements and numerical calculations of flow fields in and around Savonius rotors. J Wind Eng Ind Aerodyn. 1996;59(1):39–50.
44. Wekesa DW, Wang C, Wei Y, Zhu W. Experimental and numerical study of turbulence effect on aerodynamic performance of a small-scale vertical axis wind turbine. J Wind Eng Ind Aerodyn. 2016;157:1–14.
45. Dobrev I, Massouh F. CFD and PIV investigation of unsteady flow through Savonius wind turbine. Energy Procedia. 2011;6:711–720.
46. Kacprzak K, Liskiewicz G, Sobczak K. Numerical investigation of conventional and modified Savonius wind turbines. Renew Energy. 2013;60:578–585.
47. Sarma NK, Biswas A, Misra RD. Experimental and computational evaluation of Savonius hydrokinetic turbine for low velocity condition with comparison to Savonius wind turbine at the same input power. Energy Convers Manage. 2014;83:88–98.
48. Roy S, Saha UK. Wind tunnel experiments of a newly developed two-bladed Savonius-style wind turbine. Appl Energy. 2015;137:117–125.

Chapter 7
Viscous Flows in the Impeller Pump

Abstract Flows in the impeller pump are typical among engineering fluid flows. In this chapter, we select a low-specific-speed centrifugal pump as an example to demonstrate research methods commonly used in this aspect. This pump is equipped with long and short blades. Both the pump performance and inner flow characteristics at various flow rates are studied. The design of the pump impeller is optimized, which is expected to be validated through the proof extracted from the flow field. Unsteady numerical simulation is conducted to disclose inner flow patterns associated with the modified design. Meanwhile, an assessment of the hydraulic forces exerted on the pump components is implemented based on numerical results.

7.1 Introduction

The centrifugal pump has unparalleled advantages in a variety of applications associated with marine engineering. It adapts well to partial-load operation conditions provided that critical hydraulic components are deliberately designed. A centrifugal pump operating at low flow rates is investigated and a connection between flow patterns and acoustic parameters is established [1]. The adverse effect of low-flow-rate operation can also be testified by those worn or even damaged hydraulic components, rotary or stationary. The correspondence between unstable pump operation and low flow rate has been reinforced with consistently enriched knowledge of complex turbulent flows in the centrifugal pump. This argument is substantiated through exploring transient flows in a centrifugal pump [2].

For the low-specific-speed centrifugal pump, friction loss is remarkable due to large impeller outlet diameter. Furthermore, as this branch of centrifugal pumps operate at low flow rates, secondary flows and flow separation occur not only at impeller inlet and outlet, but also in blade passages, as is demonstrated by numerical results obtained in [3]. Apart from that, positive slope segment of pump head curve, a salient sign of operation instability, has been frequently observed in practical applications. In this context, to increase blade number but not to invite

C. Kang et al., *Methods for Solving Complex Problems in Fluids Engineering*,
https://doi.org/10.1007/978-981-13-2649-3_7

excessive blade crowding, short blades are introduced. Fundamental function of short blades has been numerically analyzed in [4]. Usually, an impeller equipped with long and short blades is termed as the compound impeller, which is schematically shown in Fig. 7.1. Also displayed in Fig. 7.1 are velocity triangles at the inlet and outlet of an individual blade passage. It is reasonable that the intrusion of the short blade helps suppress the diffusion of the flow passage between the two neighboring long blades and restrict relative flows in the blade passage. Furthermore, short blades alter essentially flow patterns in the blade passage. In addition, owing to shortened inlet part, the short blade in some extent reduces the inlet shock loss which is particularly significant in consideration of the narrow impeller passage of the low-specific-speed centrifugal pump.

It is concluded in [5] that streamwise relative velocity distribution near the short blade differs from its counterpart near the long blade. Thus, the deviation of the short blade from its original position impacts immediately velocity and pressure distributions in the impeller passage. Such a relationship is complicated under off-design conditions. Studies in this respect have rarely been reported hitherto. The effect of short blades has also been connected with cavitation phenomenon. The initial cavitation stage is studied and it is confirmed that the area of the cavity attached to the short blade is small compared to that near the long blade leading edge [6].

In view of instantaneous flows between the impeller and a radial diffuser captured in [7], the combination of a compound impeller with a radial diffuser will foster diversified flow structures and fluctuations. Not only blade and vane numbers but also the dimensions of the gap between the impeller and the diffuser have profound effects on impeller-diffuser interaction. Until now, few conclusions have been reached to offer a systematic guidance for the design of the radial diffuser. Meanwhile, in the presence of the three-dimensional gap flows, current flow measurement techniques are obviously deficient [8]. In this context, experimental work has been performed but the disparity between the test model and the commercial product should be minimized[9].

Emphasis of this chapter is placed upon a single-stage low-specific-speed centrifugal pump operating at low flow rates. Regarding the pump hydraulic

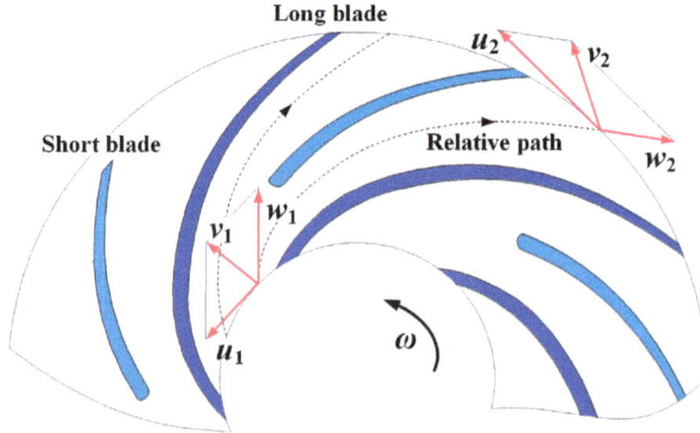

Fig. 7.1 Schematic of the impeller with long and short blades

components, a radial diffuser is deployed immediately downstream of a compound impeller, and then a spiral volute is responsible for the conversion of kinetic energy into pressure energy. The purpose of this chapter is to explore the effects of short blades on pump performance and inner flow patterns at low flow rates. Unsteady numerical simulation based upon Reynolds averaged Navier-Stokes (RANS) equations serves as a primary instrument in view of limitations of available flow measurement techniques. Three aspects, namely pump performance, flow patterns in the impeller-diffuser gap, hydraulic forces exerted upon the impeller, will be treated inter-relatedly. A further understanding of the application of the CFD technique will help to optimize the design of major hydraulic components of the pump.

7.2 Pump Structure and Parameters

The pump devoted to the current study is designed with the nominal flow rate q_V of 170 m^3/h, pump head H of 280 m. The rated rotational speed of the impeller, ω, is 5700 rpm. The specific speed of the pump can be calculated by:

$$N_s = \frac{n\sqrt{q_V}}{H^{0.75}} \quad (7.1)$$

Eq. (7.1)yields a pump specific speed value of 18. In addition, in parallel with the practical system requirements, this pump operates frequently at flow rates lower than $0.4q_V$. Therefore stable operation of the pump should be granted the highest priority.

In terms of the design of the compound impeller, there is no explicit theoretical principle that can be relied on to calculate the basic geometrical parameters. Therefore empiricism is predominant during the design process. After the impeller scheme is determined, the radial diffuser is designed following the widely applied area-ratio principle [10]. For the pump considered here, the impeller is composed of five long curved blades and five short curved blades, and the radial diffuser is composed of seven two-dimensional vanes. Both the hydraulic components and corresponding flow domains are shown in Fig. 7.2.

As per the general design methodology, the short blade of the impeller is deployed in the middle of the two adjacent long blades, as marked by 'Original design' in Fig. 7.3. As an explorative attempt, the short blades are uniformly shifted towards corresponding blade back surfaces. On one hand, working medium accommodated in the two flow passages separated by the short blade becomes more equivalent. On the other hand, at low flow rates, the development of the large vortex between the working surface of the short blade and the back surface of the long blade will thereby be restricted. Position modification for short blades has recently been adopted in [11] for the study of an inducer and the attention is concentrated on cavitation phenomenon in the inducer involving short blades. Furthermore, the

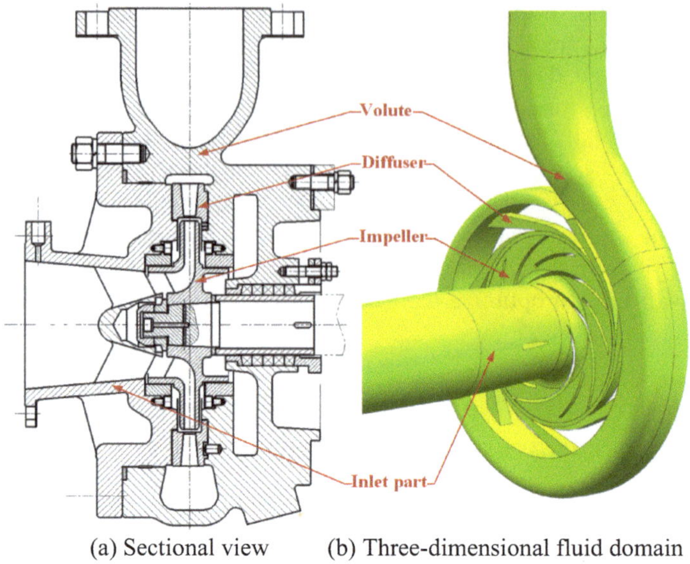

Fig. 7.2 The model pump

throat area of the diffuser channel is decreased by 10% relative to the original dimension, as is realized through reducing both the diffuser channel width and height. The influence of geometrical parameters of the gap on pump performance has been discussed in [12]. In agreement with the conclusions obtained in [12], the impeller-diffuser gap is modified through reducing the basic circle diameter of the radial diffuser from 256 to 250 mm in the present study.

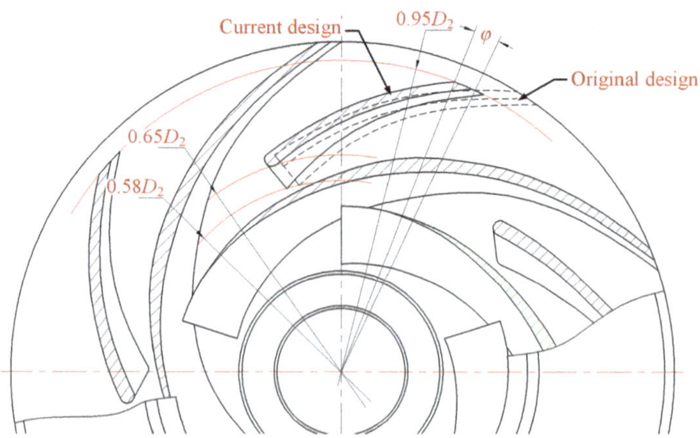

Fig. 7.3 Comparison of the original and modified impellers

7.3 Numerical Preparations

7.3.1 Governing Equations and Turbulence Model

The flows are assumed incompressible and RANS-based flow-governing equations are used. Furthermore, the practical system is sufficiently spacious to absorb the heat generated due to the friction between the working medium and the walls, so energy equation is not included in the governing equations.

For the turbulence model, we use the standard k-ε turbulence model for steady flow simulation. The preliminarily obtained steady flow field solutions are used as initial conditions for the subsequent unsteady simulations. Here, unsteady simulation employs renormalization group (RNG) k-ε turbulence model. The feasibility of this model has been confirmed in the simulation of inner flows of a centrifugal pump [13]. The time step is defined as the time span with which the impeller rotates by one degree. The convergence criteria are uniformly set to be 10^{-4}, after which the computation proceeds to the next time step. Commercial CFD code ANSYS CFX is used as the solver of the governing equations.

7.3.2 Boundary Conditions

Pure water of 22 °C serves as the working medium transported in the pump. As for the pump in practical service, a constant static pressure of 0.2 MPa is exerted on the pump inlet. This condition is fully implemented in the numerical work and thereby, cavitation phenomenon is not considered. Hence, the inlet pressure serves as the reference pressure for the calculation of pump head.

Velocity inlet boundary condition is set at the inlet of the whole computational domain. Therefore, volume flow rate can be adjusted through its dependence on inlet velocity. Flow rates ranging from 5 to 170 m^3/h are examined. Since accurate static pressure as well as velocity distributions at the domain outlet is initially unknown, outflow boundary condition is set at the outlet section. Except the parameter of static pressure, all the other parameters are associated with zero normal gradients at the outlet section of the computational domain. Such a boundary condition is particularly suitable for incompressible flow and velocity inlet boundary condition. A medium turbulence intensity of 5% is set at the inlet of the computational domain and no-slip boundary conditions are imposed on all solid walls. All surfaces wetted by the fluid are associated with surface roughness of 0.025 mm, as is parallel to surface roughness of the prototype pump. Near-wall regions are treated with scalable wall function, which is desirable relative to standard wall function in not only adapting to various $y+$ values but also facilitating grid refinement.

7.3.3 Verification of Numerical Settings

Here, prior to the determination of the final grid scheme, five sets of grids involving unstructured grids deployed in the impeller passages and structured grids discretizing the rest sub-domains are utilized to discretize the same computational domain. As a most fundamental requirement, grid quality is guaranteed for the five grid schemes. Instead of monitoring the variation of pump head with grid number, static pressure and velocity at three points deployed at the impeller eye (Point 1), in the blade channel (Point 2) and the convergent section of the volute (Point 3), are monitored at nominal flow rate, as shown in Fig. 7.4a. Static pressure and velocity at these three points are time averaged for each grid number scheme. The results are plotted in Fig. 7.4b.

Regarding the point at the impeller eye, variations of both static pressure and velocity with grid number are negligible, as shown in Fig. 7.4. Nevertheless, in the blade channel, variations of static pressure and velocity are easily identifiable, particularly with the small grid numbers. This point is subject to the effects of flow structures of various scales. Therefore, the grid dependence examination carried out on this point is saliently pertinent. In contrast, the point in the diffuser outlet is featured by high static pressure but small velocity magnitudes. With the two schemes with large grid numbers, the relative difference of static pressure or velocity is lower than 1%. In this context, the fourth grid number scheme with a set of grids totaled 8,163,260 is sufficiently reliable as far as RANS based numerical simulation is concerned.

According to conclusions obtained in [14], for unsteady simulation, the time step corresponding to impeller rotation angle of 1° allows the solver to capture

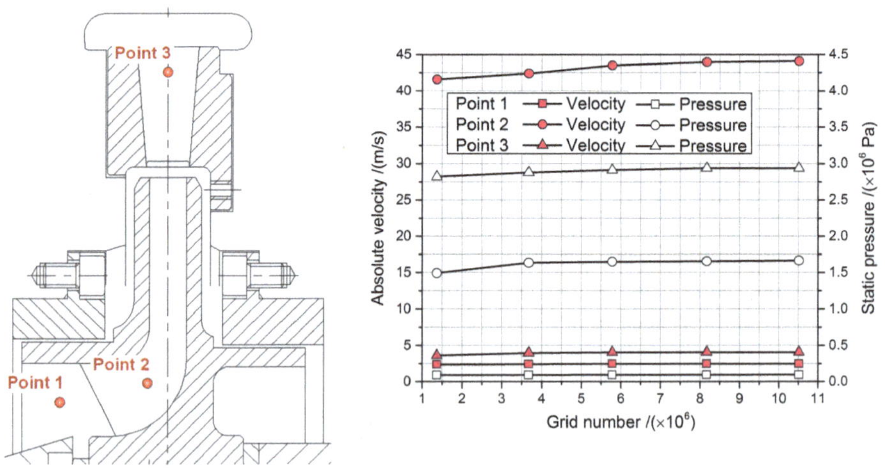

(a) Monitored points (b) Variation of flow quantities with flow rate

Fig. 7.4 Variation of static pressure and velocity with grid number at monitored points

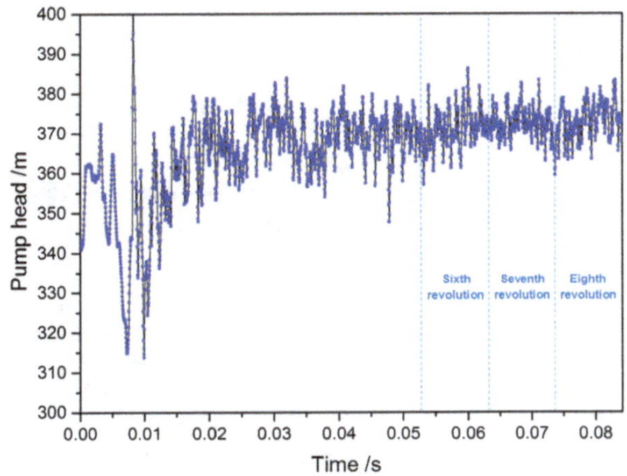

Fig. 7.5 Time-dependent pump head variation at nominal flow rate

unsteadiness of the flows. In view of the start-up instability of the unsteady simulation, totally eight impeller revolutions are accomplished and only the data associated with the eighth revolution are taken into account in post-processing and the following discussion. Figure 7.5 shows the time-dependent variation of pump head at nominal flow rate. It is seen that the statistical periodicity is accomplished after five revolutions and for the sixth, seventh and eighth revolutions, the periodicity is retained.

7.4 Results and Discussion

7.4.1 Pump Performance

A comparison is performed between the original and current schemes in Fig. 7.6 which contains four pump head curves, two of which are recorded in experiments and the other two are numerically obtained head curves based upon transiently obtained total pressure values during the eighth revolution of the impeller.

The experiments are conducted in a closed water loop, as shown in Fig. 4.17. Such an experimental rig conforms to ISO 9906:2012, which stipulates the permissible fluctuation amplitudes for critical parameters such as flow rate, rotational speed and pump head. The overall uncertainty of the pump head measurement is ±2%, pump efficiency 3.8%, with both systematic and random uncertainties considered.

For the original scheme, both experimental and numerical results show positive slope segment, alternatively, some kind of instability, as the flow rate varies from 5 to 90 m^3/h. In this case, as the flow rate rises, the increment of pump head exceeds the energy actually needed by the system, therefore the flow rate tends to increase further. With the modified design, it is seen in Fig. 7.6 that the pump head curve shape is changed significantly, particularly with the elimination of positive slope. It

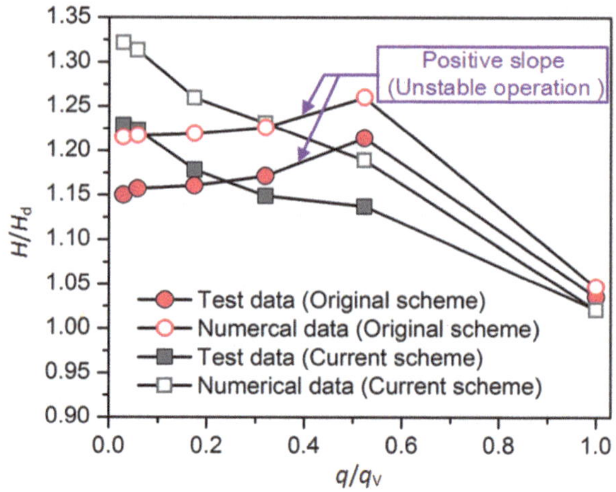

Fig. 7.6 Pump performance curves obtained experimentally and numerically

can be concluded that the inflow condition is improved with the modification of short blades at low flow rates. Meanwhile, the suppression of secondary flows at the impeller outlet is another factor alleviating the positive slope. In this connection, experimental data and numerical results are in good accordance. Furthermore, in the whole flow rate range considered, numerically obtained pump head values are consistently higher than their experimental counterparts. This is ascribed to the neglecting of volumetric loss, the friction between pump components and the underestimated friction between fluid and solid walls. Such a difference is especially apparent at low flow rates [15]. At nominal flow rate, numerical results and test data are in good agreement for the modified design.

7.4.2 Flow Patterns

In one full period of impeller rotation, impeller blades alternatively pass through some certain circumferential position residing in the gap between the impeller and the radial diffuser. Regarding the impeller-diffuser interaction, it is determined synthetically by five rotating long blades, five rotating short blades and seven stationary diffuser vanes. Without quantitative reference, such an issue is anticipated to be deciphered with instantaneous flow parameter distributions in conjunction with flow parameter fluctuations. Here, a global view of flow velocity distributions is presented firstly. Five consecutive absolute velocity distributions with the impeller rotation over the mid-span plane of the flow passages are extracted, as shown in Fig. 7.7. The three columns are associated with three low flow rates, namely 55, 35 and 15 m^3/h, respectively. In each column, time intervals between every two neighboring contours are identically defined as $T/30$, where T is the impeller rotation cycle. The whole column is designated to express the process during which a short blade, marked with black colour, passes by an individual diffuser channel. The three columns share the same initial relative position between the marked short blade and the diffuser channel surrounded by Vane A and Vane B, as denoted in Fig. 7.7.

Although each diffuser channel is featured by different relative positions between impeller blades and diffuser vanes, flow patterns in different diffuser channels are similar and are asynchronous with impeller rotation, as is perceivable in Fig. 7.7. Here, the diffuser channel between Vanes A and B is deemed as representative. At the diffuser throat, large velocity magnitude arises due to the abruptly narrowed flow space, and immediately downstream, flows apparently meander towards the spiral volute. As the flow rate decreases, velocity gradients near the diffuser throat are enhanced, and high velocity gradients are maintained as the impeller rotates. Cutting of the short blade outlet section virtually changes the profile of the gap between the

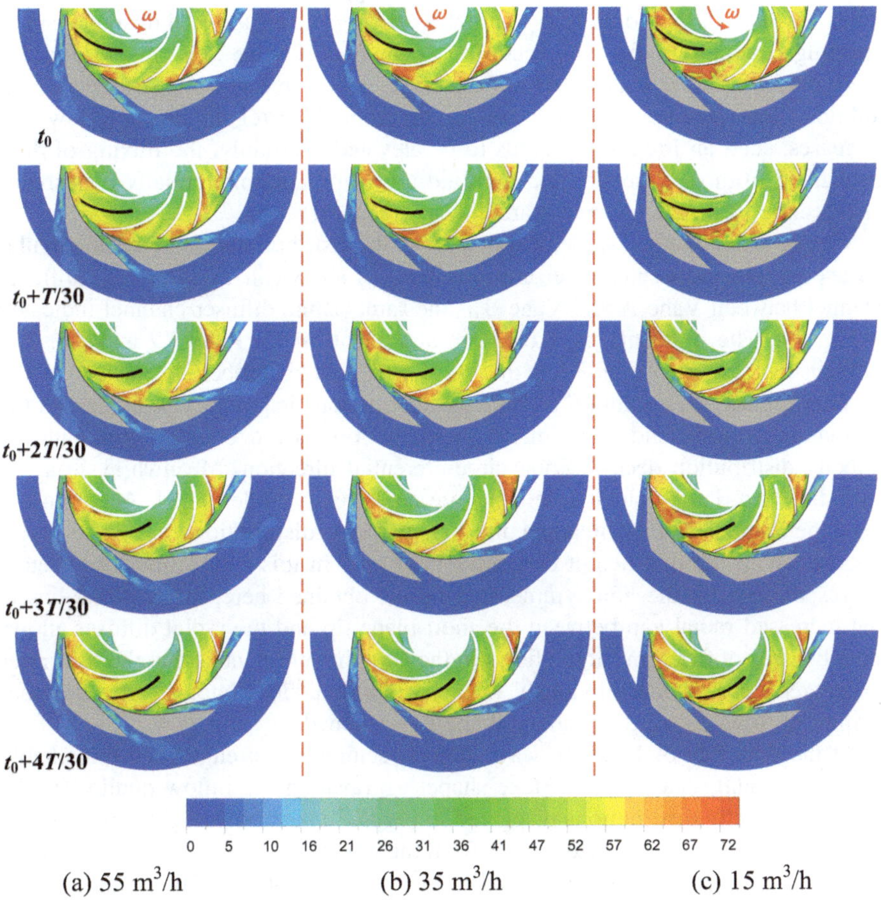

Fig. 7.7 Gap flow patterns at mid-span plane of flow passages

impeller and the diffuser as well as the periodicity of impeller-diffuser interaction. As the short blade approaches the diffuser channel, inflow quality for the diffuser channel is improved since that enlarged radial distance between the diffuser and the impeller allows the rapid adjustment of local flow angles towards favorable inflow conditions for the diffuser. Furthermore, at the flow rate of 15 m^3/h, there is a high-velocity zone at impeller outlet and near the long blade working surface and the velocity is higher than those of the other two corresponding zones at 55 and 35 m^3/h. Such a high-velocity zone nurtures flow structures of various scales at impeller outlet and has a profound effect on downstream flows in the diffuser channel.

With a further examination of flow patterns at the flow rate of 15 m^3/h, it can be found that a large portion of high-velocity fluid at impeller outlet travels along circumferential direction rather than into the diffuser passage. As a consequence, the frequencies of flows entering the radial diffuser passages are altered, forecasting the inception of rotating stall which has been described in [16]. As for fans, it has been reported that low-frequency components, approximately 3 Hz, might emerge under low flow rate conditions. Nevertheless, there is no established data relevant to centrifugal pumps. In this connection, the numerical scheme is incapable of accurately detecting those characteristic frequencies lower than the shaft frequency. In addition, flows at impeller outlet are circumferentially irregular. As the flow rate decreases, such an irregularity tends to be elevated. Certainly, the mixing of fluid discharged from the impeller and the fluid travelling circumferentially contributes significantly to flow unevenness around the impeller.

Viewed from another angle, absolute velocity distributions at the basic circular surface of one diffuser channel are unfolded, as shown in Fig. 7.8. The diffuser channel between Vane A and Vane B is the same as the diffuser channel indicated in Fig. 7.7. The moments monitored are also identical for Figs. 7.7 and 7.8. The most distinct feature of Fig. 7.8 is that no effect of blade number is detected. As the impeller rotates, the evolution of velocity distributions is not synchronous with the alternation of long and short blades. As the flow rate decreases, continuity of velocity distribution decays along circumferential direction. Meanwhile, in spanwise direction, large velocity elements are consistently undermined. According to the experimental results obtained in [17], velocity distributions are rather symmetrical about the mid-height circumferential line. In this connection, two factors are responsible for the non-symmetrical results obtained here, the first one lies in that increased radial gap between the short blade tip and the radial diffuser allows more degrees of freedom for the flows in the gap; the other one is that the flow rates considered here deviate obviously from the nominal flow rate, thereby the conformity of flows to solid boundary shape is weakened.

At the flow rate of 15 m^3/h, wavy flow structures are strengthened remarkably and many small-scale elements take shape, aggravating the inflow quality for the diffuser. Velocity distributions displayed in Fig. 7.8c bear no straightforward effect of impeller rotation, as is inseparable from the effects of short blades on the flows surrounding the impeller. Relative to the other two flow rate conditions, such a quite low flow rate is dominated by the gap flow elements which are homogenized over the circumferential direction.

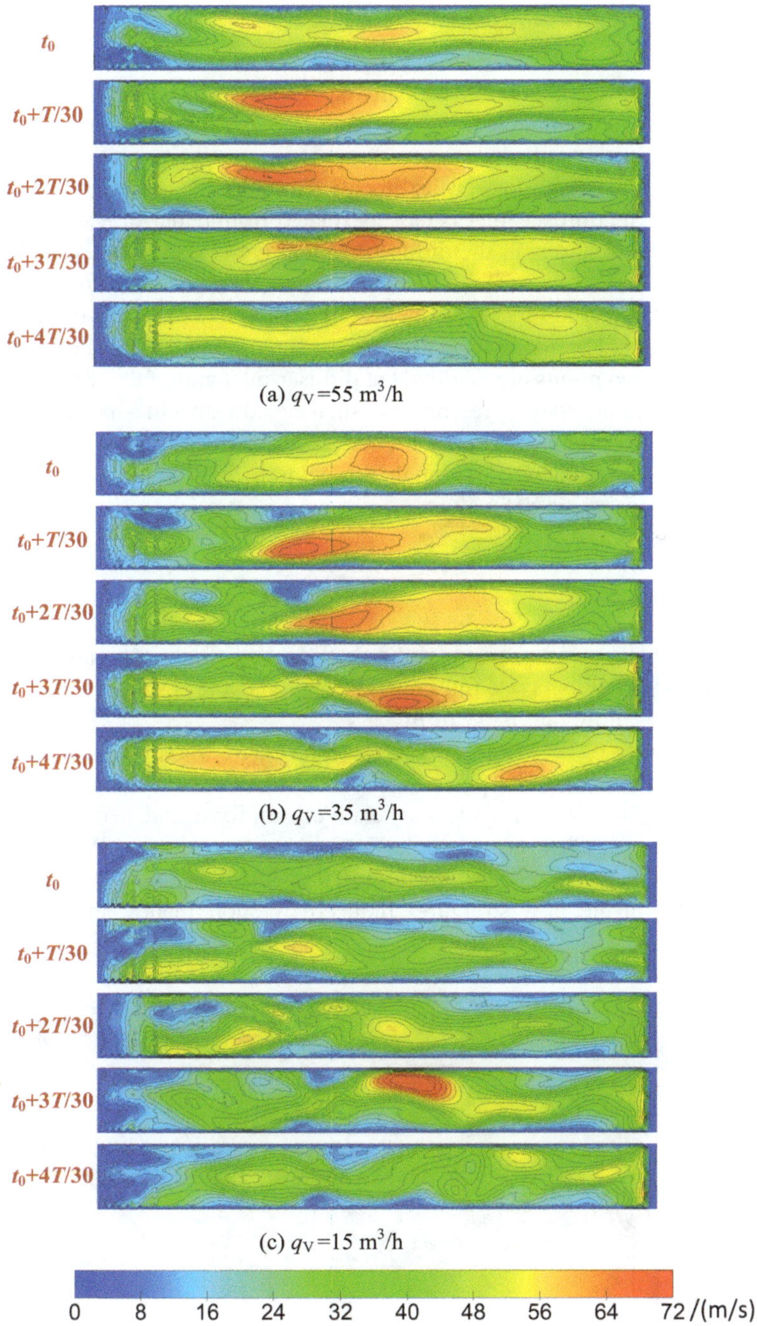

Fig. 7.8 Unfolded velocity distributions between the impeller and the diffuser

7.4.3 Pressure Fluctuations

A global view of velocity distribution in the gap between the impeller and the diffuser has been presented in Figs. 7.7 and 7.8. As for the pump considered, periodic flow behavior still deserves a comparative analysis at diffuser inlet and inside the diffuser channel, two critical locations where complex flow phenomena are anticipated. The diffuser inlet is closely related to impeller-diffuser interaction, while velocity distributions in the diffuser channel are sensitive to the variation of flow rate. Local velocity distributions in a low-specific-speed centrifugal pump is discussed in [18]. According to the conclusions obtained in [18], a two-dimensional frame can be counted on to describe flow phenomena in the pump. Therefore, at the mid-span plane, two points are deployed at diffuser inlet and in the diffuser channel to monitor transient static pressure in such a low-specific-speed centrifugal impeller. In addition, periodic characteristics reflecting the influence of short blades are anticipated to be discerned through this approach. A non-dimensional static pressure coefficient, c_p, is defined as:

$$c_p = \frac{p}{0.5 \cdot \rho u_2^2} \tag{7.2}$$

where p is the static pressure and ρ is the fluid density. Transient static pressure at the two monitored points in one full impeller rotation cycle is shown in Fig. 7.9 and the corresponding flow rate is 55 m^3/h.

At the flow rate of 55 m^3/h, the two static pressure curves deviate from each other considerably. At p_1, the point deployed at the diffuser inlet, static pressure values over the full rotation cycle are relatively uniform and periodicity is well maintained although there are multiple dominating frequencies. Point p_1 carries apparent information associated with the rotating impeller, and those small peaks are produced with the flow structures induced by short blades. In contrast, low

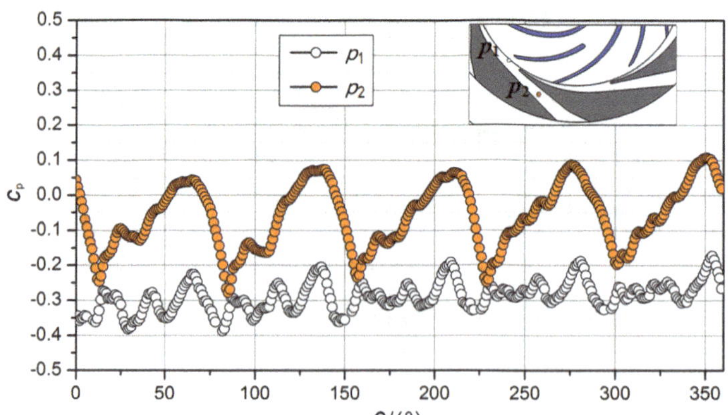

Fig. 7.9 Variation of static pressure with impeller rotation at q_v = 55 m^3/h

peaks are shielded remarkably by five salient peaks at p_2, and overall pressure magnitudes are large relative to their counterparts at p_1.

As the flow rate decreases from 55 to 35 m^3/h, the spatial periodicity of static pressure variation attenuates evidently, as shown in Fig. 7.10. In this connection, not just p_1 but also p_2 are characterized by irregular time evolution of static pressure. As the flow rate decreases, the deficiency of fluid becomes evident with respect to the flow passage, so flow parameter oscillation, as well as flow structure deformation, occurs both at impeller outlet and in the diffuser channel. Meanwhile, secondary flows and vortices are locally excited, which is in agreement with the numerical results obtained at different flow rates [19]. As a consequence, static pressure variations at both the two points exhibit weak periodicity.

As the flow rate is further reduced to 15 m^3/h, as shown in Fig. 7.11, the irregularity is intensified and pressure fluctuations in circumferential direction are obvious. It thereby opens the possibility of the prosperity of low-frequency components. There are two issues that current numerical simulation cannot address. One is that the frequency components lower than shaft frequency cannot be distilled from the transient data due to the limitation in temporal resolution. The other one lies in the lack of quantitative information which incurs the difficulty in building the relationship between local flow pattern and flow-induced fluctuations. It is noticeable that at such a quite low flow rate, the two static pressure curves are approaching each other in terms of both magnitude and periodicity, and some mechanism connected with the strong deficit in the flow rate pervades the two monitored locations.

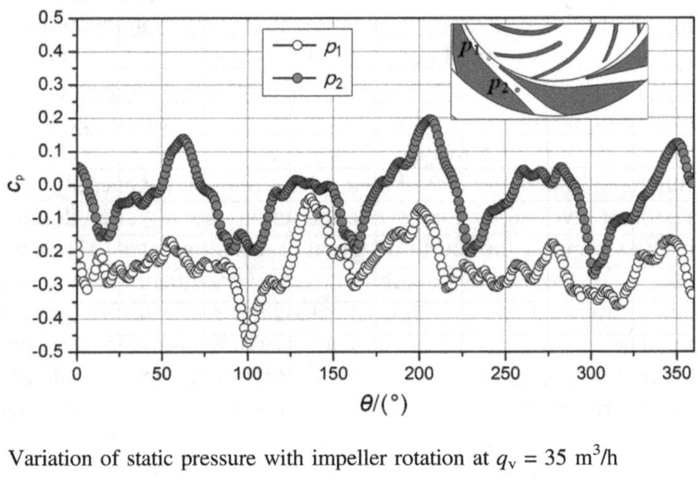

Fig. 7.10 Variation of static pressure with impeller rotation at q_v = 35 m^3/h

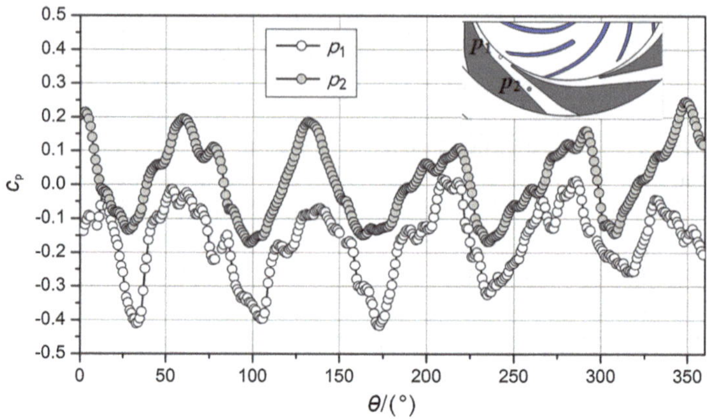

Fig. 7.11 Variation of static pressure with impeller rotation at q_v = 15 m^3/h

7.5 Hydraulic Forces Exerted on Impeller Blades

Hydraulic forces serve as an immediate metric of the rotational stability of the impeller. Special attention is devoted to hydraulic forces under off-design conditions for an ordinary centrifugal pump and it was found that hydraulic forces are rather high at low flow rates [20]. In this context, it should be pointed out that hydraulic forces exerted upon general centrifugal impellers have been studied extensively and fruitfully. Nevertheless, the compound impeller has not been discussed in this aspect. With unsteady numerical results, the integral of transient static pressure distributions over blade surfaces enables the acquisition of resultant hydraulic forces. In Fig. 7.12, at the same three flow rates aforementioned, both the resultant hydraulic forces exerted upon the whole impeller and the short blades are presented. As the impeller rotates, the orbit of resultant hydraulic force develops with the increase of the marked number. For instance, the orbit progresses along the route of 1 → 2→3 → 4 → 5 in Fig. 7.12a.

For a centrifugal impeller enclosed by a spiral volute, it is generally believed that orbits of resultant hydraulics forces exerted upon the impeller dominate overwhelmingly the second or the fourth quadrant at low flow rates [21]. Nevertheless, in Fig. 7.12a, c, e, the one-cycle orbits cover every quadrant in a rather even fashion. This is ascribed to the employment of the radial diffuser instead of the spiral volute. The hydraulic force component in y direction increases considerably as the flow rate decreases from 55 to 35 m^3/h. Regarding the flow rate of 15 m^3/h, both hydraulic force components in x and y directions are apparently large compared to their counterparts associated with the other two flow rates.

From q_V = 55 to 35 m^3/h, there is an increase in the magnitude of the hydraulic force acting upon the whole impeller. In contrast, the hydraulic forces exerted upon the short blades vary only slightly and meanwhile, the two orbit patterns displayed in Fig. 7.12b, d are similar. In terms of overall hydraulic force magnitude, the

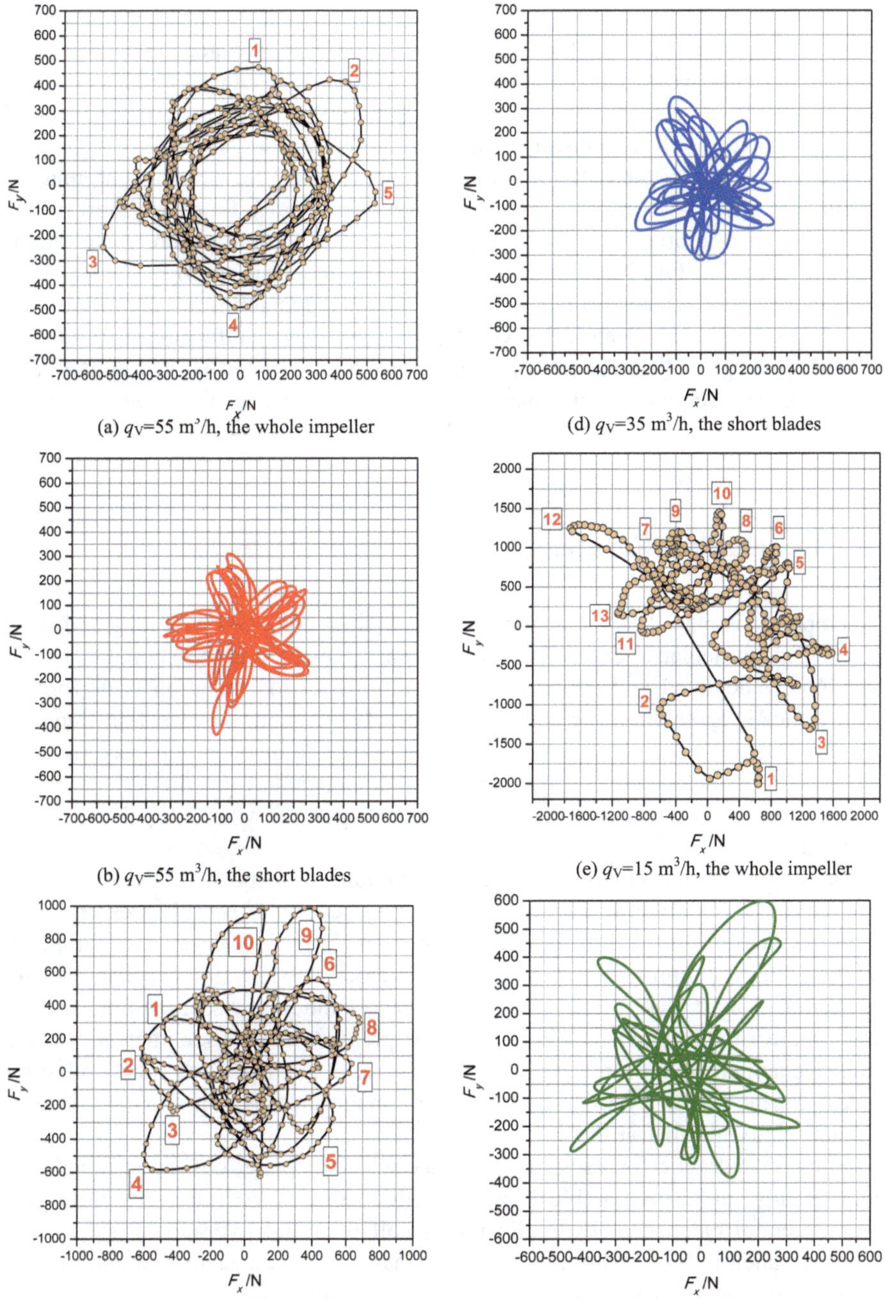

(a) q_V=55 m³/h, the whole impeller

(d) q_V=35 m³/h, the short blades

(b) q_V=55 m³/h, the short blades

(e) q_V=15 m³/h, the whole impeller

(c) q_V=35 m³/h, the whole impeller

(f) q_V=15 m³/h, the short blades

Fig. 7.12 Orbit diagram of resultant hydraulic forces acting upon the whole impeller and the short blades contribution of short blades is small relative to that of the long blades. At q_v = 15 m^3/h, the orderliness of the hydraulic force orbit associated with the whole impeller is rather weak and circumferentially distributed bulges cannot constitute even an approximate circle. Nevertheless, variation of the hydraulic forces exerted upon short blades still demonstrates clear circumferential regularity.

With the cutting of the outlet section of the short blade, flow patterns around the impeller are altered accordingly, as can be seen in Fig. 7.13. Since the short blades bear small surface area relative to long blades, small hydraulic forces are anticipated. In addition, Fig. 7.13 indicates similar pressure distributions over working and back surfaces of the short blades, as also proves the insignificant contribution to pump head. At low flow rates, the working medium is not uniformly distributed in the blade passages. Nevertheless, the transient hydraulic forces exerted on the short blades remain circumferentially balanced, as is shared by all the three flow rates considered.

7.6 Concluding Remarks

The emphasis is placed on a low-specific-speed pump equipped with long and short blades, and particular attention is placed upon the effects as the short blades are intentionally deviated and cut as well. A multi-aspect investigation is performed using unsteady simulation, which is uniquely advantageous in revealing transient flow field details. Thereby, numerical results throw light on pump performance, velocity distributions between the compound impeller and the diffuser, transient pressure variations and hydraulic forces exerted upon the whole impeller and the short blades.

The scheme of modified short blades is immune from positive slope of pump head curve at low flow rates. In essence, the modification of short blades improves the inflow quality for the diffuser at low flow rates. In the gap between the impeller and the diffuser, circumferential uniformity of velocity distribution is enhanced with the shortened short blades. However, blade effects cannot be discerned immediately downstream of the impeller. Transient static pressure at diffuser inlet and in the diffuser channel proves the spatial periodicity is associated with long and short blades. Furthermore, overall static pressure magnitude is high in the diffuser channel relative to that at the diffuser inlet.

Orbits of hydraulic forces acting upon the whole impeller cover four quadrants rather evenly, as is inseparable from the radial diffuser. At quite low flow rate, circumferential balance of hydraulic forces collapses and hydraulic force magnitudes are saliently high. Although with low magnitudes, transient hydraulic forces exerted upon the short blades cover all the four quadrants.

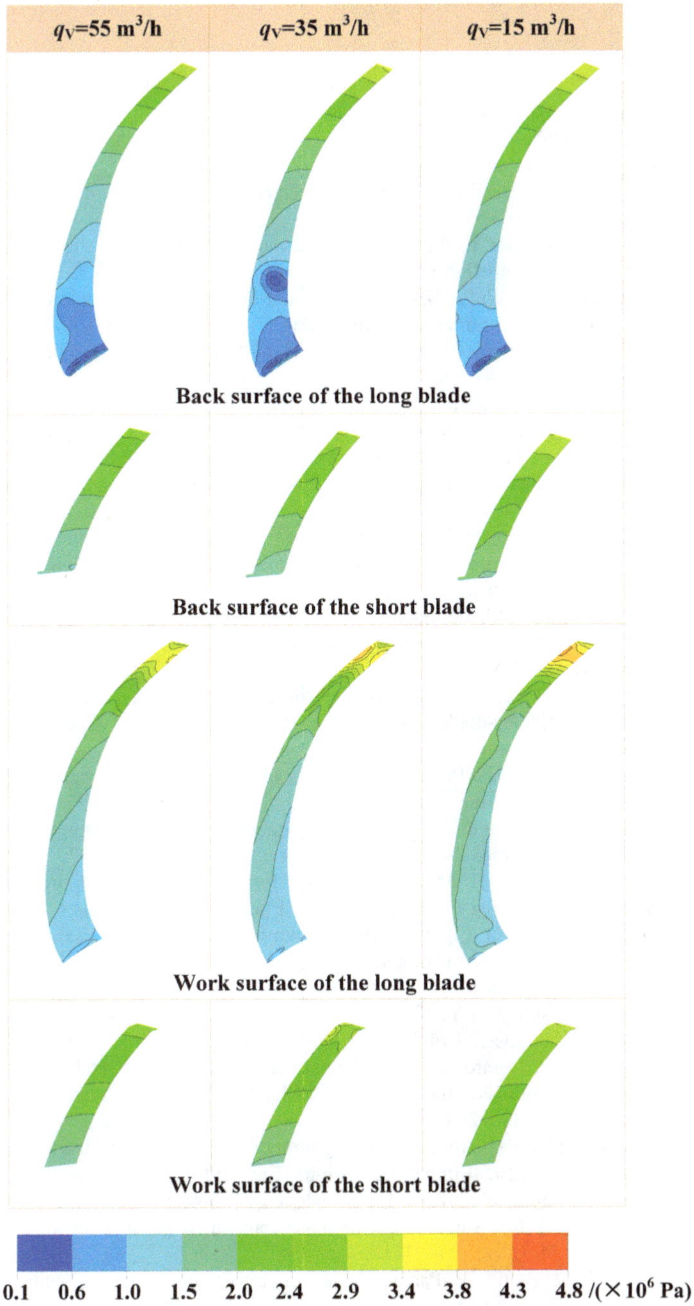

Fig. 7.13 Pressure distributions over the blade surfaces at different flow rates

References

1. Huang J, Geng S, Wu R, Nie C, Zhang H. Comparison of noise characteristics in centrifugal pumps with different types of impellers. Acta Acust. 2010;35(2):113–118.
2. Feng J, Benra F-K, Dohmen HJ. Application of different turbulence models in unsteady flow simulations of a radial diffuser pump. Forsch Ing. 2010;74(3):123–133.
3. Cui B, Lin Y, Jin Y. Numerical simulation of flow in centrifugal pump with complex impeller. J Therm Sci. 2011;20(1):47–52.
4. Yuan S, Zhang J, Yuan J, He Y, Yuedeng F. Effects of splitter blades on the law of inner flow within centrifugal pump impeller. Chin J Mech Eng. 2007;20(5):59–63.
5. Yang W, Xiao R, Wang F, Wu Y. Influence of splitter blades on the cavitation performance of a double suction centrifugal pump. Adv Mech Eng. 2014;2014(1):963197.
6. Thai Q, Lee C. The cavitation behavior with short length blades in centrifugal pump. J Mech Sci Technol. 2010;24(10):2007–2016.
7. Petit O, Nilsson H. Numerical investigations of unsteady flow in a centrifugal pump with a vaned diffuser. Int J Rotating Mach. 2013.
8. P Dupont, G Caignaert, G Bois, T Schneider, Rotor-stator interactions in a vaned diffuser radial flow pump. In: Proceedings of ASME fluids engineering division summer meeting; 2005. p. 1087–1094.
9. Feng J, Benra F-K, Dohmen HJ. Numerical study on impeller-diffuser interactions with radial gap variation. In: Proceedings of 4th WSEAS international conference on fluid mechanics and aerodynamics, Elounda, Greece; 2006. p. 289–294.
10. Anderson HH. Prediction of head, quantity and efficiency in pumps—the area ratio principle. In: The 22nd annual fluids engineering conference, the American society of mechanical engineers, New Orleans, LA, Mar 9–13; 1980.
11. Guo X, Zhu Z, Cui B, Li Y. Effects of the short blade locations on the anti-cavitation performance of the splitter-bladed inducer and the pump. Chin J Chem Eng. 2015;23 (7):1095–1101.
12. Ozturk A, Aydin K, Sahin B, Pinarbasi A. Effect of impeller-diffuser radial gap ratio in a centrifugal pump. J Sci Ind Res. 2009;68:203–213.
13. Kang C, Li Y. The effect of twin volutes on the flow and radial hydraulic force production in a submersible centrifugal pump. Proc IMechE Part A. 2015;229(2):221–237.
14. Petit O, Bosioc AI, Nilsson H, Muntean S, Susan-Resiga RF. Unsteady simulations of the flow in a swirl generator using OpenFOAM. Int J Fluid Mach Syst. 2008;4:199–208.
15. Guo X, Zhu Z, Cui B, Shi G. Effects of the number of inducer blades on the anti-cavitation characteristics and external performance of a centrifugal pump. J Mech Sci Technol. 2016;30:3173–3181.
16. Sano T, Yoshida Y, Tsujimoto Y, Nakamura Y, Matsushima T. Numerical study of rotating stall in a pump vaned diffuser. J Fluids Eng. 2002;124(2):363–370.
17. Abdelmadjid A, Saad B, Gerard B, Patrick D. Numerical and experimental comparison of the vaned diffuser interaction inside the impeller velocity field of a centrifugal pump. Sci China Technol Sci. 2011;54(2):286–294.
18. Hesse NH, Howard JHG. Experimental investigation of blade loading effects at design flow in rotating passages of centrifugal impellers. J Fluids Eng. 1999;121(4):813–823.
19. Stel H, Amaral GDL, Negrão COR, Chiva S, Estevam V, Morales REM. Numerical analysis of the fluid flow in the first stage of a two-stage centrifugal pump with a vaned diffuser. J Fluids Eng. 2013;135(7):071104–071109.
20. Barrio R, Fernández J, Blanco E, Parrondo J. Estimation of radial load in centrifugal pumps using computational fluid dynamics. Eur J Mech-B/Fluids. 2011;30:316–324.
21. Gülich JF. Centrifugal pumps. Berlin: Springer; 2010.

Chapter 8
Cavitation in the Condensate Pump

Abstract Emphasis of this chapter is placed on the cavitation phenomenon arising in a condensate pump. The condensate pump is unique since the amount of gas in the inflow cannot be neglected. It is the responsibility of optimal design to relieve the cavitation effect. Research tools are to be used to validate the innovation. For the pump considered, the impeller meridional profile is modified to improve the cavitation performance. Furthermore, shortened blades are employed to minimize the disturbance of cavitation on blade pressure distribution. Proposed impeller schemes are examined using experimental and numerical methods. Experiments enable the evaluation of pump head and critical net positive suction head. Cavitating flows in the pump are virtually visualized using the CFD technique. The relationship between flows and pump performance is anticipated to be established based on the obtained results.

8.1 Introduction

Vaporization and condensation between liquid and vapor phases reflect the most phenomenal essence of cavitation. Regarding the condensate pump, it is typically deployed downstream of the condenser and delivers liquid medium to the heater and then the deaerator [1]. Because the temperature of the medium that enters in the condensate pump is fairly high, the risk of cavitation is clearly evident. As cavitation occurs, the blockage of pump blade channels by cavitation bubbles might result in the deficiency in water supply to the downstream facilities; the consequence is disastrous. Therefore, the cavitation study is of distinct significance for the condensate pump, as compared to the pumps used in common industrial applications.

Cavitation visualization is a straightforward and popular experimental strategy for investigating cavitation phenomenon [2]. As a prerequisite, the light used to illuminate cavities or cavitation bubbles must penetrate into the flow field. Thus an elaborately fabricated pump model is necessitated to ensure model transparency as well as imaging fidelity. This inevitably gives rise to the inconsistency between the

C. Kang et al., *Methods for Solving Complex Problems in Fluids Engineering*,
https://doi.org/10.1007/978-981-13-2649-3_8

experimental model and the prototype. So far, cavitation visualization has been practiced overwhelmingly in pumps with low specific speed. Otherwise, cavitation can be studied using the CFD technique. Based on CFD results, the inception and evolution of cavitation in the pump can be demonstrated virtually. It should be noted that for the cavitating flows in engineering, advanced flow simulation methods such as large eddy simulation (LES) are demanding and difficult to implement [3].

In this chapter, to improve the cavitation performance of a centrifugal condensate pump, the geometry of the impeller is modified and the impeller eye area is enlarged. Furthermore, half of the impeller blades are shortened to minimize the blade surface area invaded by cavitation. The geometric modification is similar to that recorded in Chap.7 but the purposes are different. The validation of the new impeller schemes is substantiated using experimental and numerical methods. Experiments enable the evaluation of the cavitation performance of the pump considered, and numerical work is conducted to locate cavities in the pump and to describe cavity geometry. The influence of the flow rate and pump inlet pressure on cavity shape and cavitation volume fraction is examined. The resistance to cavitation with the introduction of the proposed impellers is anticipated to be assessed through both external and internal behavior of the pump. Moreover, the similarities and difference between the presented impeller schemes are clarified through a comprehensive comparison.

8.2 Cavitation Feature of the Centrifugal Pump

As continuous suction of liquid is achieved using the centrifugal pump, the static pressure declines gradually from the pump inlet pipe to the impeller eye. It is estimated that the static pressure reaches its minimum near the blade inlet edge, thereby facilitating the local occurrence of cavitation. Images acquired using high speed photography (HSP) are displayed in Fig. 8.1 to illustrate typical cavitation phenomenon in the centrifugal impeller. In Fig. 8.1a, relatively weak cavitation is indicated. For each blade, cavities are attached to the suction surface and near the blade inlet edge. It is observed that the cavitation patterns are not identical for each blade. As cavitation is intensified, as shown in Fig. 8.1b, cavities extend apparently into the blade passages; meanwhile, the cavitation volume fraction is increased. In some blade passages, cavities might block the inlet part, which depends on the blade geometry and the blade number. Therefore, even for the centrifugal pump exclusively, cavitation patterns can be diverse.

Cavitation in the centrifugal pump is subject to the effect of multiple factors. In this context, pump operation parameters such as the flow rate and impeller rotational speed are influential. As the flow rate deviates from the design flow rate, the direction of the relative velocity component at blade inlet edge might be altered considerably, and subsequently, cavities might be shifted from the blade suction surface to pressure surface [4]. Certainly, the blade geometry also assumes an important role in this case. Hitherto, bubble cavitation and attached cavitation have

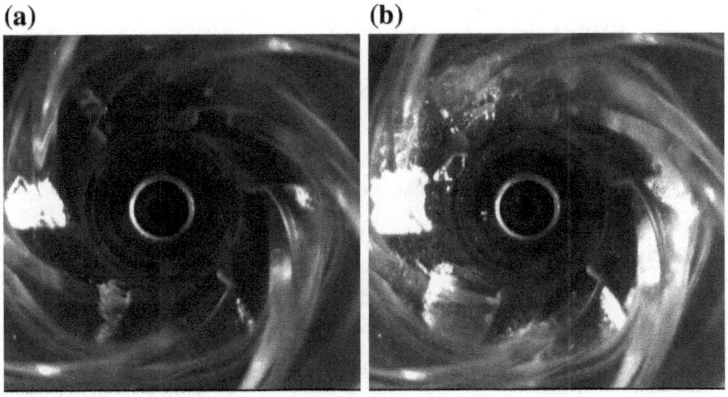

Fig. 8.1 Images of cavitation in a centrifugal pump impeller

been acknowledged in the centrifugal pump. However, as for the impetus behind the distinct cavitation patterns, viewpoints are diverse. In most occasions, cavitation is determined by multiple factors and it is difficult to distinguish the contribution of any individual factor.

8.3 Modification of Impeller Geometry

The impeller considered is vertically installed and the bulk flow enters the impeller with downward velocity, as depicted in Fig. 8.2. Upstream of the impeller eye, the medium motion is driven by dual impetuses, pressure gradients and gravity. In the impeller eye, the influence of rotating blades is predominant and the medium begins to be transported radially towards impeller outlet. Until now, such an impeller orientation has rarely been reported. Based on common design principles, the meridional profile of the impeller denoted by the dashed lines in Fig. 8.2 is selected. Here, both the front and rear shrouds are intentionally expanded, as indicated with red solid lines in Fig. 8.2. Thus an enlarged impeller eye is presented. With such a scheme, the relative velocity at blade inlet will be reduced and thereby local static pressure increased. Meanwhile, cavities are expected to be confined near the rear shroud in the impeller eye and do not invade the flow in blade channels.

Based on the modified meridional profile shown in Fig. 8.2, blades of equal length are devised, as shown in Fig. 8.3a. Blade inlet edge is the most direct part of sensing the effect of cavitation. Thus some blades are shortened at the inlet part and the blade length is reduced, as exhibited in Fig. 8.3b. The purpose is to minimize the influence of cavities on the flow parameter distribution over blade surface. Meanwhile, to ensure sufficient pump head, more blades are employed. In this chapter, two impellers will be investigated for the same condensate pump, one impeller involves enlarged impeller eye and blades of equal length, the other impeller is associated with enlarged impeller eye and long and short blades.

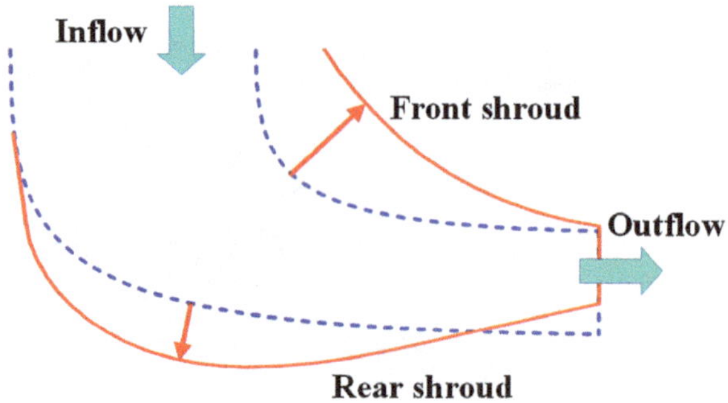

Fig. 8.2 Common and modified meridional profiles of impeller

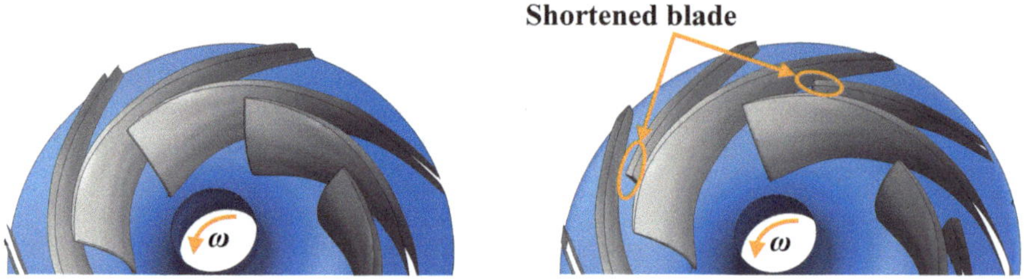

(a) Impeller with blades of equal length (b) Impeller with long and short blade

Fig. 8.3 Schematic view of two impeller schemes

8.4 Pump Performance Experiment

The design flow rate of the condensate pump, $q_{V0,}$ is 180 m^3/h, the nominal rotational speed, n, is 1480 rpm, and the design pump head, H_0, is 32 m. Two impellers, one is equipped with seven blades of equal length and the other with five long and five short blades, are devoted to the experiment. The two impellers bear identical impeller outlet diameter, D_2, of 350 mm, and identical impeller inlet diameter, D_1, of 215 mm.

The pump performance experiment is conducted in a closed hydraulic loop which conforms to the ISO 9906:2012 standard [5]. A schematic configuration of the loop is shown in Fig. 8.4. The pump is installed vertically and connected with adjacent pipes via horizontal inlet and outlet pipes. In consideration of the large capacity of the water tank as well as sufficient length of the pipes, the temperature rise during the experiment is rather low, and the water temperature is found to remain at (26 ± 2)°C. To induce cavitation, the vacuum pump was adjusted while the valves mounted upstream of the pump are kept at the fully open state. Such a practice prevents the cavitation promotion in the pump due to cavitation bubbles in the incoming flow. These bubbles are created near the valve plate due to locally narrowed flow section.

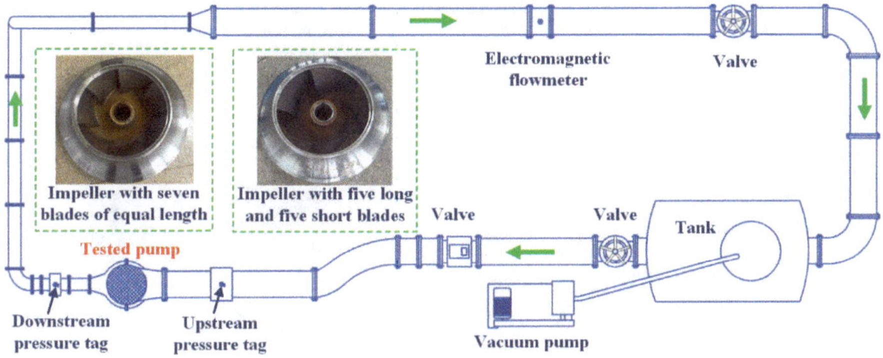

Fig. 8.4 Schematic view of the test loop and images of the two impellers

Based on experimental results, the pump head and the critical net positive suction head ($NPSH_c$) are plotted in Fig. 8.5 as a function of the flow rate. Regarding the two impeller schemes, pump head decreases inversely and consistently with the flow rate. In comparison, the pump head of the seven-bladed impeller is higher than its counterpart, as applies to all flow rate conditions. The critical net positive suction head is obtained as the pump head drops by 3%. It is salient that at the design flow rate q_{V0}, very small values of $NPSH_c$ are observed. The impeller with long and short blades is responsible for the minimum $NPSH_c$ of 0.76 m. Such an $NPSH_c$ value is rather low as far as centrifugal pumps are concerned. Within the flow rate range of 0.5–1.2q_{V0}, favorable cavitation performance is well kept for the impeller with long and short blades, and at 1.3 and 1.4q_{V0}, cavitation performance is undermined remarkably. As for the impeller with seven blades of equal length, the cavitation performance is slightly low compared to that of the impeller with blades of unequal length. The rise of $NPSH_c$ from 1.2 to 1.4q_{V0} agrees with that of its counterpart. At 0.5q_{V0}, the deterioration of cavitation performance deserves much attention. Normally, for the centrifugal pump, cavitation performance is elevated gradually with the decrease of flow rate.

Both the pump head and cavitation performance associated with the two proposed impellers are disclosed explicitly with experiment. Of interest are the cavitation behavior in the pump and how cavitation affects the external pump performance [6]. In view of the condensate pump structure, it is not an easy task to describe the internal flow field with non-intrusive measurement and visualization techniques [7]. In this context, unsteady numerical simulation is performed to describe the detailed cavity geometry and cavitation evolution in the pump. Moreover, the influence of the flow rate and the pump inlet pressure on cavitation is anticipated to be clarified.

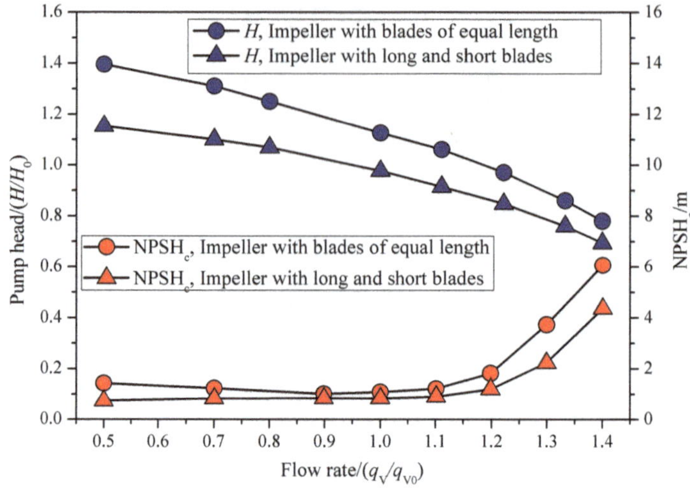

Fig. 8.5 Variation of pump head and $NPSH_c$ with flow rate

8.5 Numerical Preparations

8.5.1 Governing Equations, Turbulence Model and Cavitation Model

In parallel with the experimental preparation, pure water of 26 °C is used as the flow medium in the computational domain. Three-dimensional incompressible flows are assumed and the flows are governed by Reynolds averaged Navier-Stokes (RANS) equations. Here, the shear stress transport (SST) k–ω turbulent model is selected. The merits of this turbulence model have been explained in Chap. 2. The cavitation model of Zwart-Gerber-Belamri (ZGB model) is adopted. This model has also been introduced in Chap. 2.

8.5.2 Grid Deployment Scheme

A three-dimensional geometric model of the condensate pump is constructed with the dimensions identical with those of the prototype. Hence the computational domain was obtained by means of Boolean processing and local geometry adjustment. The domain accommodating the impeller equipped with long and short blades is displayed in Fig. 8.6. Prior to the operation of spatial discretization, an inlet segment and an outlet segment are deployed upstream and downstream of the computational domain, respectively. The lengths of the two segments are $5 \times D_{inlet}$ and $5\times D_{outlet}$, respectively. D_{inlet} and D_{outlet} represent the diameters of the pump

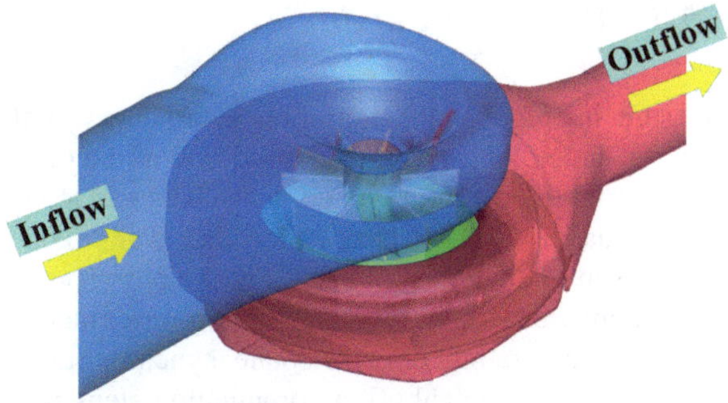

Fig. 8.6 Schematic diagram of the computational domain for the impeller with long and short blades

inlet and outlet pipes, respectively. The extension of pump inlet part ensures that flows upstream of the pump are fully developed, and the auxiliary outlet segment helps to prevent the backward effect of the flow at the outlet section on the flows inside.

The computational domain exhibited in Fig. 8.6 is discretized with approximately 3.8 million unstructured tetrahedral grids, as is attained using the commercial mesh-generation code of ICEM CFD. Local grids are illustrated in Fig. 8.7. The grids near the impeller blade surface are refined to capture those near-wall flow phenomena that are exposed to high velocity gradients [8]. The equiangle skewness is 0.22. The values of y^+ range from 27 to 67. For the interface between rotating and non-rotating subdomains, the fidelity of data exchange is ensured through adjusting grid density near the interface. Regarding the grid deployment for the computational domain associated with the pump equipped with the seven-bladed impeller, the grid number is about 3.6 million, and similar strategies of grid construction are utilized.

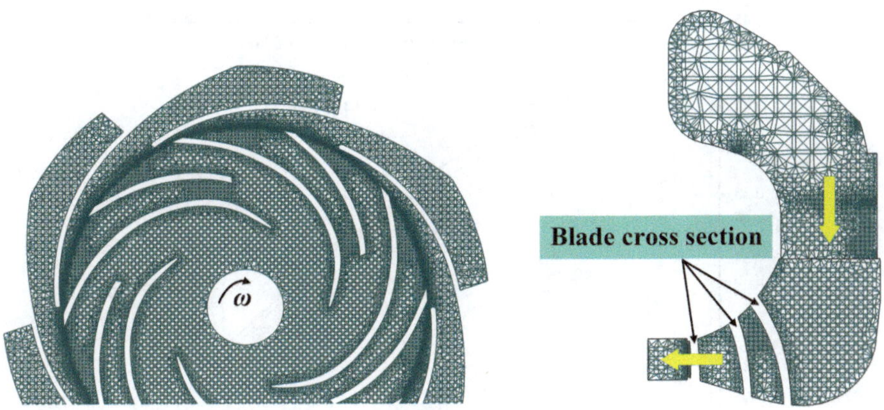

Fig. 8.7 Cross-sectional grid deployment for the impeller scheme with long and short blades

8.5.3 Boundary Conditions

Pressure inlet boundary condition is assigned at the inlet of the computational domain, and velocity boundary condition at the outlet. No-slip boundary condition is set for all the medium-wetted walls. The surface roughness of the impeller blade is set equal to 0.025 mm and the other walls 0.05 mm. The boundary condition setting is in line with the operation of the pump prototype, as ensures the credibility of the numerical results [9]. The sliding mesh method is used to guarantee that the grids in the rotating domain rotate synchronously with the impeller. The time interval between neighboring computation steps is set equal to the time span with which the impeller rotates for one degree. At each time step, the convergence threshold of 10^{-4} is designated for all monitored flow quantities.

8.5.4 Validation of the Computational Scheme

During the unsteady simulation, the pump head is monitored instantaneously as the impeller rotates. After seven rotation periods, periodic variation of numerical results is attained. In this context, pump head data acquired in the eighth and ninth periods of impeller rotation are time-averaged and the resultant pump head is used in subsequent comparison. These procedures are performed at each flow rate. For the impeller with long and short blades, the deviation of numerically obtained pump head with respect to the experimental data in Fig. 8.5 is described in Fig. 8.8. Apart from that, at n = 980 rpm, the pump head is measured and corresponding numerical work is performed as well. Flow patterns in the pump depend on the rotational

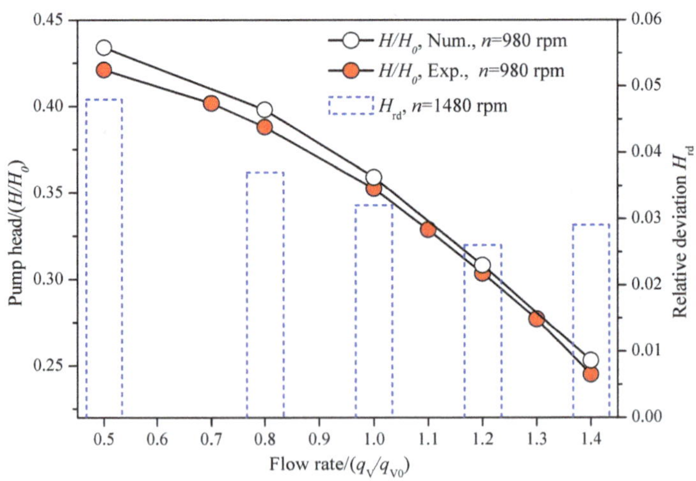

Fig. 8.8 Comparison between numerical (Num.) and experimental (Exp.) results. H_{rd} denotes the relative deviation of pump head at n = 1480 rpm

speed, as extends the capability of the numerical strategy in distinguishing flow structures and in predicting the effect of the internal flow on pump performance.

As indicated in Fig. 8.8, at nominal rotational speed, the minimum deviation between the experimental and numerical results arises at 1.2 q_V, as is reasonable since the volumetric loss are not taken into account in the numerical simulation. Regarding the impeller considered, it is appreciable that a small amount of medium cannot be discharged into the blade channels due to the enlarged impeller eye. Large flow rate enhances the agreement between numerical and experimental situations. Meanwhile, the largest relative deviation emerges at 0.5 q_V and is less than 5%. At n = 980 rpm, the gap between numerical and experimental results is further narrowed relative to the situations at n = 1480 rpm. Moreover, the variation of the deviation with the flow rate is analogous for the two rotational speeds. As the rotational speed is low, the constraint effect of solid boundary on flow patterns is more remarkable. Thus the numerical results are closer to the experimental data. Such a tendency proves the credibility of the numerical strategy adopted.

8.6 Numerical Results and Discussion

8.6.1 Geometric Features of Cavitation

General description of cavitation phenomenon puts an emphasis on the position where cavitation occurs together with the cavity profile, both of which can be obtained based on numerical results [10]. Here, for the impeller with blades of equal length, at available net positive suction head ($NPSH_a$) of 2.0 m, instantaneous cavities in the impeller passage are presented in Fig. 8.9. The iso-surfaces in Fig. 8.9 correspond to the cavitation volume fraction of 0.5. It is observed in Fig. 8.9a that the predominant cavitation type is the attached cavitation at blade suction surface, as conforms to the common knowledge about the cavitation phenomenon in the centrifugal pump. A salient cavity pattern is the cavity sinking, as marked in Fig. 8.9a. The cavity size tends to minimize as the cavity extends in circumferential direction. Such a cavity shape is compatible with the modified profile of the rear shroud. Meanwhile, this well reflects the joint effect of the rotating blades and local flow pattern on cavitation [11]. Cavities at each blade inlet are different in terms of the area covered by the cavity. Some cavities stretch towards blade outlet and occupy a considerable percentage of the area between the two neighboring blades.

The time interval between the two images displayed in Fig. 8.9a, b is one full rotation period. Although the two impeller rotation positions are identical, the non-similarity in cavity pattern is evident. In Fig. 8.9b, small cavities sparsely distributed away from the rear shroud are observed. With the rotation of the impeller, these small cavities might enter the blade channel with the main stream.

Fig. 8.9 Cavity patterns in the impeller with blades of equal length

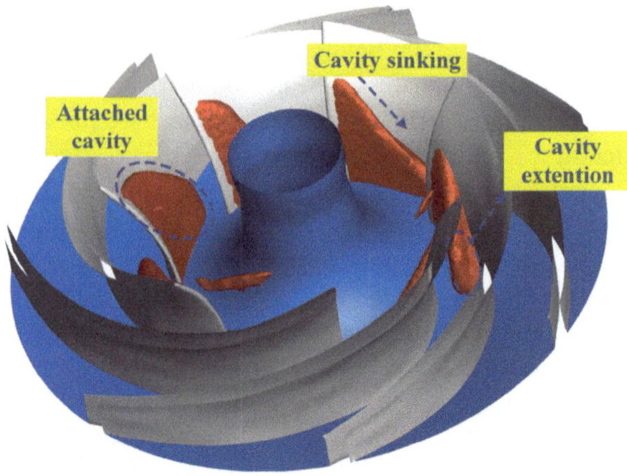

(a) t_0

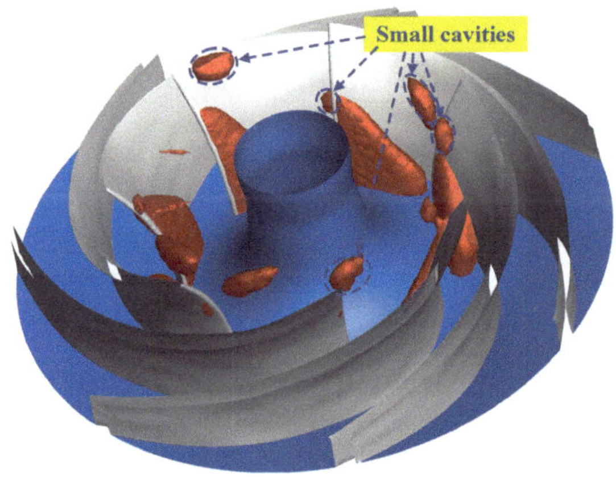

(b) t_0+T

Concurrently, the sinking and accumulated cavities near the rear shroud and blade inlet edge remain prevalent. In this context, it seems that the integrity of the large cavities is impaired and small cavities are detached from large cavities. The total cavity volume is expected to keep stable.

The cavity patterns associated with the impeller with long and short blades are shown in Fig. 8.10. The cavity profiles displayed in Fig. 8.10 are rather clear and a small total cavity volume is perceptible, as compared with those exhibited in Fig. 8.9. Similarly, attached cavities near the blade suction surface are overwhelming, and cavity sinking is representative. Some cavities tend to extend into the blade channel after separating from the blade surface, and the extension is limited relative to that indicated in Fig. 8.9. This is ascribed to the enlarged blade inlet area with shortened blades. Meanwhile, the imparity of cavity shape is observable at each long blade inlet. In Fig. 8.10a, b, individual cavity shape near each blade inlet alters considerably after a full rotation period has been consumed. The most remarkable difference between Figs. 8.9 and 8.10 lies in two aspects. One is that cavities are attached to long blades, while short blades are immune from cavity attachment. The other aspect is that drifted small cavities are absent in Fig. 8.10, and the cavity pattern is relatively monotonous.

Based on the time-dependent cavity topology exhibited in Figs. 8.9 and 8.10, it is anticipated that cavity volume fluctuation might affect the pump head. The variation of pump head for the impeller equipped with long and short blades is monitored at $1.0q_{V0}$. The results corresponding to $\mathrm{NPSH_a}$ = 3.0 m and 1.0 m are plotted in Fig. 8.11, which covers the seventh, eighth and ninth rotation period.

Fig. 8.10 Cavity patterns in the impeller with long and short blades

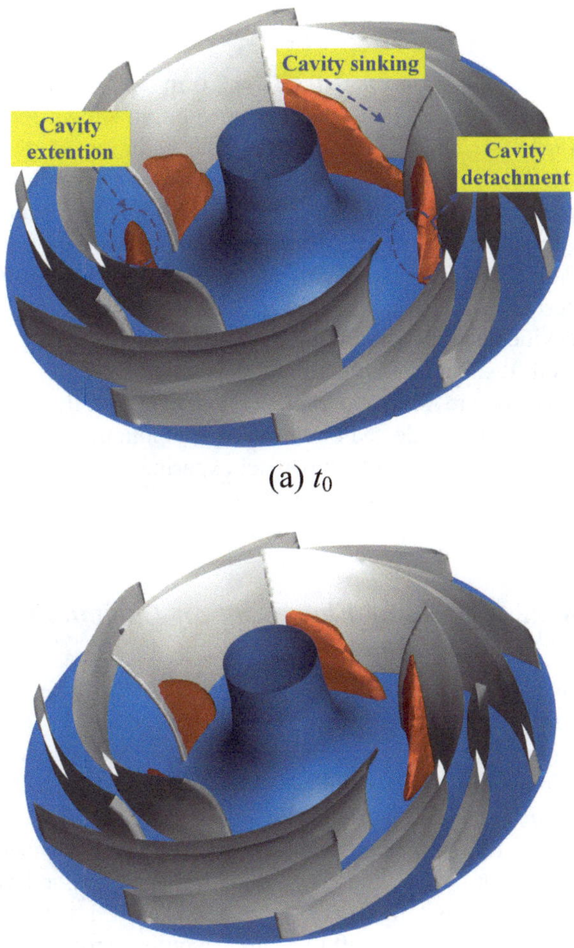

(a) t_0

(b) t_0+T

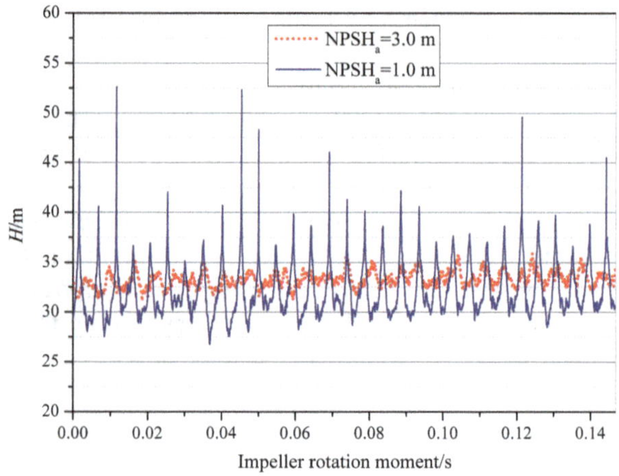

Fig. 8.11 Variation of pump head with impeller rotation

The situations at NPSH$_a$ = 3.0 m and 1.0 m can be assessed preliminarily through the cavitation performance curve shown in Fig. 8.5. In this context, with Fig. 8.11, new light is shed on the variation of the pump head under cavitation conditions. At NPSH$_a$ = 3.0 m, cavitation is undeveloped in the pump, and the pump head fluctuates slightly as the impeller rotates. In comparison, as NPSH$_a$ declines to 1.0 m, the pump head varies violently, as shown in Fig. 8.11. In this case, cavities have a significant effect on static pressure distribution over the blade surface. Moreover, the flow rate of water changes instantaneously as the extended cavities partially block the blade inlet area, evoking the variation of pump head as well. It should be noted that the time scales of cavitation evolution and pump head variation are not comparable. One depends on flow excitation and the other one is a global effect reflected on the solid components. Meanwhile, pump head fluctuations obtained numerically have no experimental correspondence since that available pump head measurement technique cannot attain such a high temporal resolution.

8.6.2 *Effect of Flow Rate on Cavitation at NPSH$_a$ = 3.0 m*

In Fig. 8.11, at NPSH$_a$ = 3.0 m and 1.0q_{V0}, the pump head fluctuates insignificantly. In this connection, cavities emerge but do not undermine the pump working capability. Typical cavity patterns are traced and presented in Fig. 8.12. For comparison, at the same NPSH$_a$, representative cavity patterns corresponding to 0.5q_{V0} and 1.3q_{V0} are displayed as well.

At 1.0q_{V0}, cavities are attached to the blade suction surface, and the cavity extension in circumferential direction is suppressed evidently. This explains why a favorable cavitation performance is achieved at the design flow rate. As the flow rate increases from 1.0 to 1.3q_{V0}, cavities expand considerably and a large portion of the surface of the long blades is covered by cavities. Meanwhile, small cavities

Fig. 8.12 Cavitation at different flow rate for impeller with long and short blades

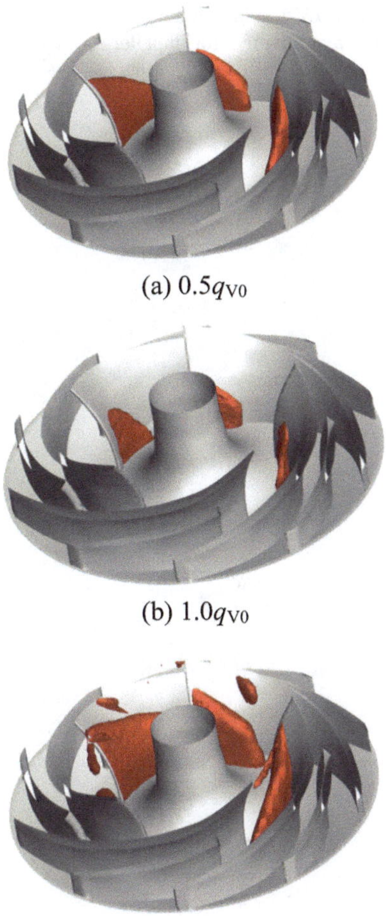

(a) $0.5q_{V0}$

(b) $1.0q_{V0}$

(c) $1.3q_{V0}$

are generated and are transported with the bulk flow, so these small cavities reinforce the cavitation effect on the pump head. The overall cavity size at $0.5q_{V0}$ is larger than that at $1.0q_{V0}$, but the cavitation performance at $0.5q_{V0}$ is still favorable. It should be noted that the attached cavities at $0.5q_{V0}$ are rather stable, and this confirms that the sinking cavities exert insignificant effect on the pump head.

In Fig. 8.12, both cavity shape and cavity volume vary with the flow rate, but the short blades are not affected by cavities. Even at $1.3q_{V0}$, the considerably extended large cavities have not reached the inlet edge of the short blades, and the cavities shrink rapidly with the downstream extension. This furnishes a sound support to the maintenance of the pump head.

For the impeller with blades of equal length, typical cavity patterns at different flow rates and $NPSH_a$ = 3.0 m are displayed in Fig. 8.13. At $1.0q_{V0}$, the cavity

Fig. 8.13 Cavitation at different flow rate for impeller with blades of equal length

(a) $0.5q_{V0}$

(b) $1.0q_{V0}$

(c) $1.3q_{V0}$

integrity near each blade inlet edge is witnessed and the cavities are confined tightly onto the blade suction surface. As the flow rate increases, cavitation development is accelerated drastically, as shown in Fig. 8.13c. Both the attached cavities and drifted cavities are prevalent. For large cavities, they have detached clearly from the blade surface and occupy a considerable portion of the blade inlet area. At $0.5q_{V0}$, although the overall cavity volume is small, cavity shape and size are diverse. With reference to the cavitation performance data shown in Fig. 8.5, the slightly lower cavitation performance at $0.5q_{V0}$ relative to that at $1.0q_{V0}$ is largely ascribed to the drifted small cavities.

It is of importance to quantify the total cavity volume and substantiate a comparison between the two impellers. At $NPSH_a$ = 3.0 m, at different flow rates, the time-dependent relative cavity volume is averaged within the ninth rotation period and the result is plotted in Fig. 8.14. The relative cavity volume is defined as the ratio of total cavity volume to the impeller chamber volume.

As the flow rate increases, the relative cavity volume increases consistently, as is shared by the two impellers. The overall relative cavity volume of the impeller with long and short blades is higher than its counterpart. As the flow rate exceeds $1.1q_{V0}$, relative cavity volume increases sharply with the flow rate. The situation in the impeller with blades of equal length is compatible with the cavity patterns shown in Fig. 8.13c. Due to the increase of the relative cavity volume, the interface between the vapor and liquid phases is enlarged, as facilitates the mass transfer between vapor and water [12]. Therefore, at high flow rates, cavitation enhances the mass transfer in the impeller but impairs the pump head. It is thereby concluded that the impeller with long and short blades is more capable of resisting cavitation.

8.6.3 *Effect of NPSH$_a$ on Cavitation at Design Flow Rate*

As a crucial boundary condition, the static pressure at pump inlet is directly related with NPSH$_a$. To depict the influence of NPSH$_a$, static pressure distributions over the blade surface and rear shroud are extracted from the numerical results. In Fig. 8.15, at NPSH$_a$ = 2.0 m, 1.5 m and 1.0 m, pressure contours for the impeller with long and short blades are presented. It is seen that low pressure prevails in the impeller eye area. Among the three cases, the decrease of NPSH$_a$ reinforces the invasion of low pressure towards the inter-blade area, as shown in Fig. 8.15. As NPSH$_a$ decreases, low pressure area in the rear shroud is enlarged clearly. Nevertheless, for the three cases shown in Fig. 8.15, the short blades are not disturbed by the diffusion of low pressure launched from the impeller eye area, as is in accordance with the cavity patterns shown in Fig. 8.9. Meanwhile, at NPSH$_a$ = 1.5 m and 1.0 m, the pressure distributions over the rear shroud imply that low pressure is prone to be homogenized in the impeller eye area.

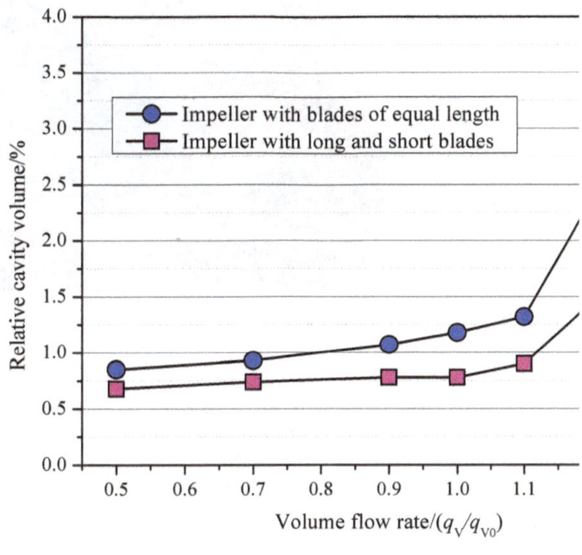

Fig. 8.14 Relative cavity volume at different flow rate at NPSH$_a$ = 3.0 m

In Fig. 8.16, pressure distributions in the impeller with blades of equal length are exhibited. Pressure gradients in the impeller eye area are apparent, as differs from that shown in Fig. 8.15. As $NPSH_a$ drops, pressure distribution in the impeller eye

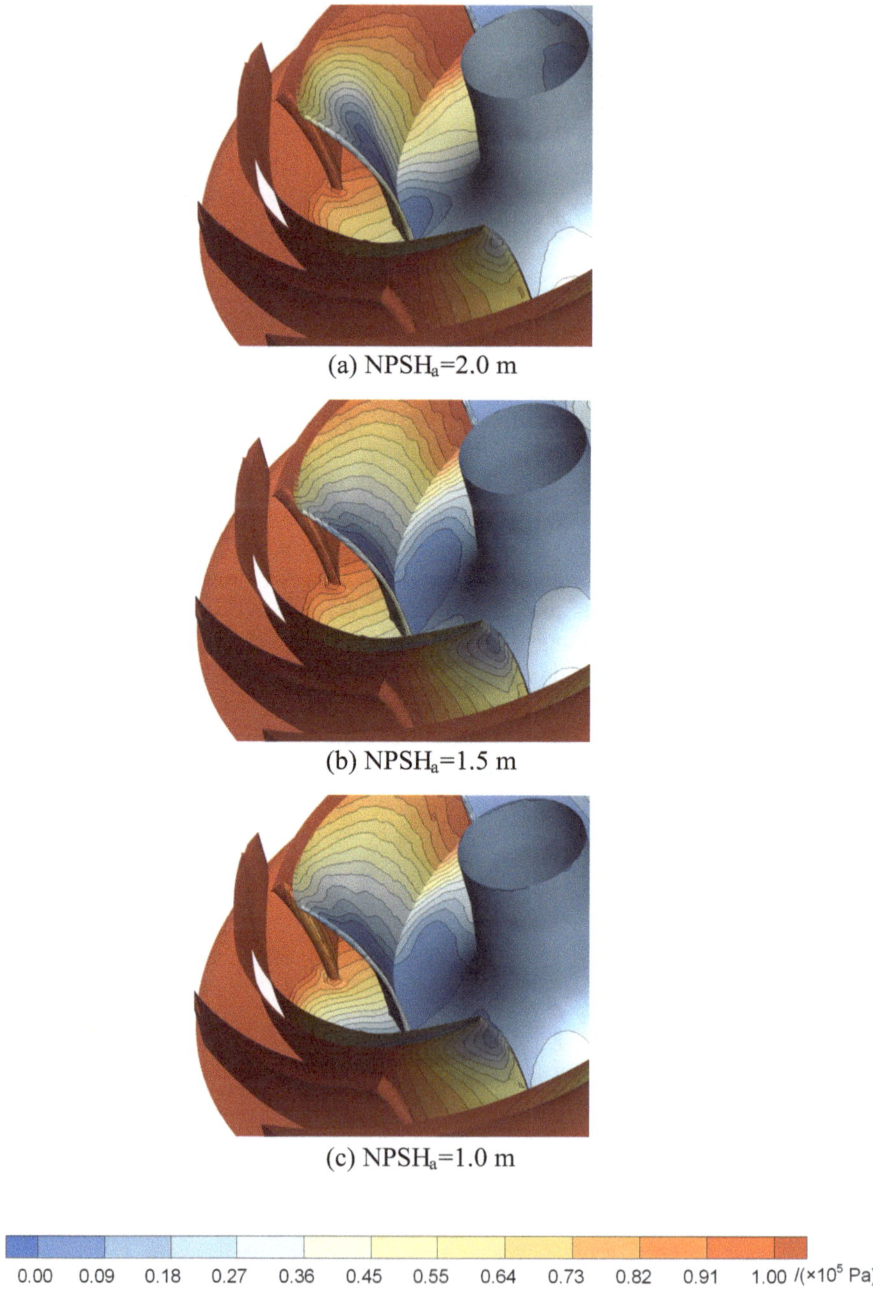

Fig. 8.15 Flow patterns for impeller with long and short blades

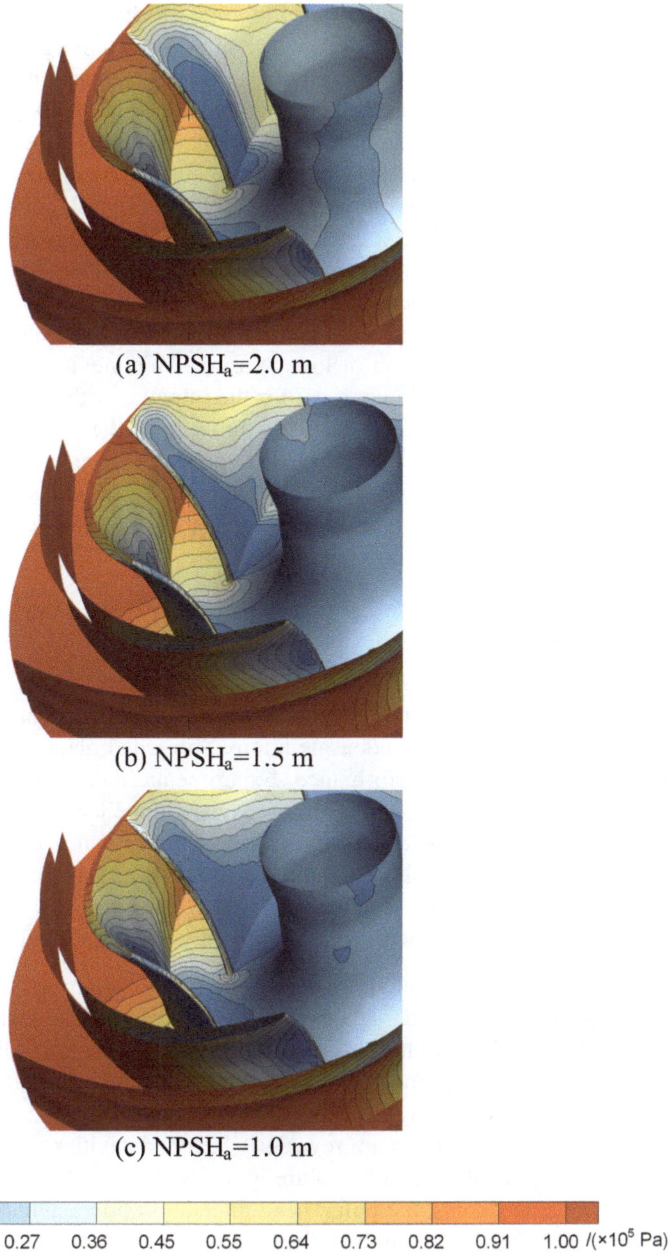

Fig. 8.16 Flow patterns for impeller with blades of equal length

area is progressively uniformized. Meanwhile, low pressure diffuses along the blade suction surface. On the surface of the blade and the rear shroud, the profiles of the area covered by very low static pressure are noticeable. At blade inlet edge, very low pressure diffuses remarkably from the rear to front shroud. In this connection, with reference to those small drifted cavities, their emergence is relevant to the diffusion of low pressure. Furthermore, near the intersection of blade suction surface and the rear shroud, the circumferential spreading of low pressure is intensified as $NPSH_a$ decreases, but the low-pressure area tends to be suppressed towards the rear shroud, as is compatible with the geometry of the attached cavities.

The advantages of the short blades are manifested through the comparison between Figs. 8.15 and 8.16. At $NPSH_a = 1.5$ m and 1.0 m, a highly vacuum state is predictable in the impeller eye. It is evident that low pressure imposes a significant effect on the blade surface. As for the impeller with blades of equal length, pressure distributions over the blade working and suction surfaces have been modified due to the invasion of low pressure. This inevitably impairs the working capability of the blades. In contrast, for the impeller with long and short blades, as shown in Fig. 8.15, although high vacuum affects the inlet parts of the five long blades, the five shortened blades are still operating under regular conditions, as can be concluded from the pressure distributions at the two sides of the shortened blades. Consequently, the loss in the pump head is minimized.

8.7 Conclusions

A special meridional impeller profile is proposed to treat the cavitation phenomenon occurring in a condensate pump. Shortened blades are adopted to further improve the cavitation performance. Experimental data demonstrate that the cavitation performance associated with the devised impellers has attained a fairly high level. To trace the root cause of cavitation performance elevation, flows in the impeller are numerically investigated. Cavity topological characteristics, as well as the relationship between cavities and static pressure distributions are clarified. Major conclusions obtained in the presented work are as follows.

(1) Critical net positive suction head of the two impellers is less than 1.0 m at the design flow rate, and high cavitation performance is achieved. The impeller with long and short blades is preferable in consideration of the consistently high cavitation performance over a wide range of flow rate. Cavitation performance deteriorates drastically as the flow rate increases further.
(2) Diverse cavity patterns are revealed. Attached cavities at blade suction surface are predominant. A salient feature is cavity sinking towards the rear shroud. As the impeller rotates, the pump head fluctuates under cavitation conditions, as is intensified as cavity volume rises. Both cavities on blade surface and drifted cavities contribute to the increase in cavity volume.
(3) Cavity volume increases monotonously with the flow rate. The impeller with long and short blades is responsible for low cavity volume. At high flow rate, the impeller with blades of equal length is characterized by the prevalence of both drifted cavities and attached cavities. The attached cavities remain predominant in the impeller with long and short blades.

(4) As available net positive suction head declines, low pressure tends to be homogenized in impeller eye. The invasion of low pressure into the blade channel is evident near the rear shroud in the impeller with blades of equal length. The pump head of the impeller with long and short blades is undermined to a limited extent since that shortened blades are not affected by cavities.

References

1. Puthiyavinayagam P, Selvaraj P, Balasubramaniyan V, Raghupathy S, Velusamy K, Devan K, Nashine BK, Padma Kumar G, Suresh kumar KV, Varatharajan S, Mohanakrishnan P, Srinivasan G, Bhaduri AK. Development of fast breeder reactor technology in India. Prog Nucl Energy. 2013;101:19–42.
2. Zhao L, Mo Z, Sun L, Xie G, Liu H, Min D, Tang J. A visualized study of the motion of individual bubbles in a venturi-type bubble generator. Prog Nucl Energy. 2017;97:74–89.
3. Pyszczek R, Kapusta ŁJ, Teodorczyk A. LES numerical study on in–injector cavitating flow. J Power Technol. 2017;97:52–60.
4. Kang C, Zhang G, Li B, Feng Y, Zhang Z. Virtual reconstruction and performance assessment of an eroded centrifugal pump impeller. Proc Inst Mech Eng, Part C: J Mech Eng Sci. 2017;231:2340–2348.
5. ISO 9906:2012, Rotodynamic pumps–Hydraulic performance acceptance tests–Grades 1, 2 and 3.
6. Iosif A, Sarbu I. Numerical modeling of cavitation characteristics and sensitivity curves for reversible hydraulic machinery. Eng Anal Boundary Elem. 2014;41:18–27.
7. Chow Y-C, Lee Y-H, Chang Y-C. Image-based measurements for examining model predictability of cavitation on a marine propeller surface. Ocean Eng. 2017;138:161–169.
8. Ahuja V, Hosangadi A, Arunajatesan S. Simulations of cavitating flows using hybrid unstructured meshes. J Fluids Eng. 2001;123:331–340.
9. Limbach P, Skoda R. Numerical and experimental analysis of cavitating flow in a low specific speed centrifugal pump with different surface roughness. J Fluids Eng. 2017;139:10120–1–101201–8.
10. Pierrat D, Gros L, Pintrand G, Le Fur B, Gyomlai Ph. Experimental and numerical investigations of leading edge cavitation in a helicon-centrifugal pump. In: The 12th international symposium on transport phenomena and dynamics of rotating machinery, Hawaii, USA, 17–22 Feb 2008.
11. Abdel-Maksoud M, Hänel D, Lantermann U. Modeling and computation of cavitation in vortical flow. Int J Heat Fluid Flow. 2010;31:1065–1074.
12. Colombet D, Goncalvès Da Silva E, Fortes-Patella R. On numerical simulation of cavitating flows under thermal regime. Int J Heat Mass Transf. 2017;105:411–428.

Chapter 9
Structural Aspect of the Impeller Pump

Abstract For fluid machinery, the interaction of fluid and solid components cannot be neglected. Sometimes, engineers care more about the solid parts than flows. Therefore, the effects of fluid on the solid parts have attracted efforts from multiple aspects. It should be noted that the shortages of current experimental techniques in treating these issues are apparent. Instead, numerical simulation plays an important role in the previously published studies. In this chapter, two cases are presented. The first case is a pump transporting high-temperature molten salt and the other case is a condensate pump. Numerical simulation is used to treat the problems with the two cases. In this connection, a joint application of CFD and FEA is substantiated. Transient pressure distributions obtained using CFD are used as the initial load exerted on the surface of the pump components. Then stress, deformation and vibration mode of the pump rotor are calculated with the finite element method. The purpose is not to dig into the flow field but to inspect the solid components. The maximum stress, stress distribution over the surface of the solid components, and the deformation are obtained. Furthermore, natural frequencies are calculated through the modal analysis.

9.1 Flow-Structure Interaction in the Molten Salt Pump

9.1.1 A Brief Introduction of the Molten-Salt Pump

The pump transporting molten salt of high temperature is usually designed as a vertical structure featured by a long pump shaft [1]. Such a pump type is unique in consideration of the applications in chemical process to transport ionic membrane caustic soda, carbonate and nitrate. Normally, the temperature of the medium involved is above 400 °C. So far, the application of such a branch of pumps has been extended into diverse chemical fields, nuclear energy utilization and thermal power exploitation [2–5]. Of significance is the pump performance, which is determined by multiple factors [6, 7]. In this context, stable and safe pump operation, instead of pump efficiency, is granted the highest priority. The rotor, which

C. Kang et al., *Methods for Solving Complex Problems in Fluids Engineering*,
https://doi.org/10.1007/978-981-13-2649-3_9

includes the impeller, pump shaft and other components rotating with the impeller, interacts with the medium and vibrates due to the fluid excitation. Furthermore, rotor vibration characteristics exert a profound effect on the operation reliability of the entire pump unit. At off-design conditions, the inner flows of the pump are irregular, and local flows foster complex flow patterns and turbulent fluctuations. As for the coupling of flow, heat and pump components, not only flow aspects but also solid pump components, are covered in current research. The strength analysis of the pump rotor for calculating stress and modal is thereby necessitated.

In 1989, Childs et al. introduced the concept of fluid-solid coupling into the field of centrifugal pump by investigating the fluid flow in the gap between the pump casing and the impeller shroud [8]. Subsequently, the enthusiasm on the method of fluid structure interaction and modal analysis in fluid machinery research has never withered [9, 10]. The fluid-solid coupling simulation for a single-bladed centrifugal pump is carried out in [11], and hydraulic excitation and displacement of the pump rotor are analyzed. Peng et al. uses finite element method to examine the strength of a large centrifugal pump, and to optimize the structure by modifying the length and thickness of the ribbed plate [12]. In [13, 14], the effect of blade thickness on vibration characteristics of a centrifugal pump is analyzed, and it is found that the variation of blade thickness has little effect on the natural frequency of the centrifugal pump itself. In [15] the first ten predominant natural frequencies of a centrifugal pump, and emphasis is laid on the resonance phenomenon which undermines pump are analyzed operation. In [16], a bearing cooling system for a molten-salt pump is designed, the air velocity and temperature distributions are analyzed, and thermal-structure interaction is substantiated. The modal analysis of a large pump turbine is conducted in [17] and the damage to the rotor due to vibration is manifested as the excitation frequency is closed to the natural frequency of the rotor. In [18], the fluid-solid coupling of a centrifugal pump is studied, and the influence of unsteady pressure pulsation on the rotor structure is described. The heat-flow coupling method has been employed to investigate the temperature field and thermal stress distribution over the two-dimensional turbine blade [19]. Langthjem et al. study the flow-induced noise in a centrifugal pump based on a two-dimensional assumption and conclude that the coupling between the rotating blade and medium serves as the main source of noise [20].

9.1.2 *Numerical Set-Up for Flow Simulation*

9.1.2.1 Geometrical Model

The nominal flow rate of the high temperature molten salt pump model is 200 m^3/h and the pump head of 65 m. The rotational speed is 1450 r/min. The centrifugal impeller is equipped with six blades. The flow computational domain is composed of inlet, impeller, volute and outlet parts. As for the structure computation domain,

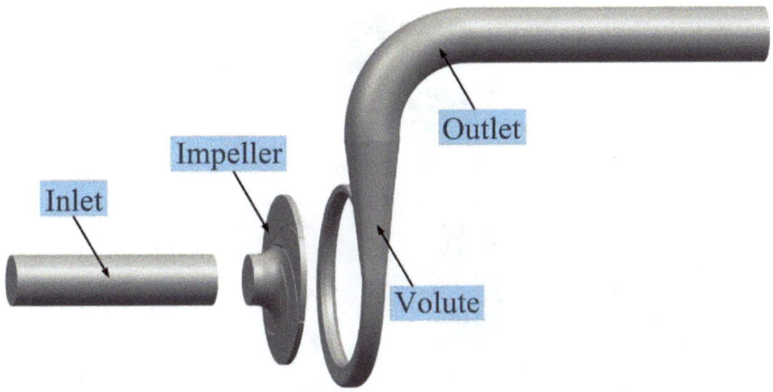

Fig. 9.1 Flow domain of the molten salt pump

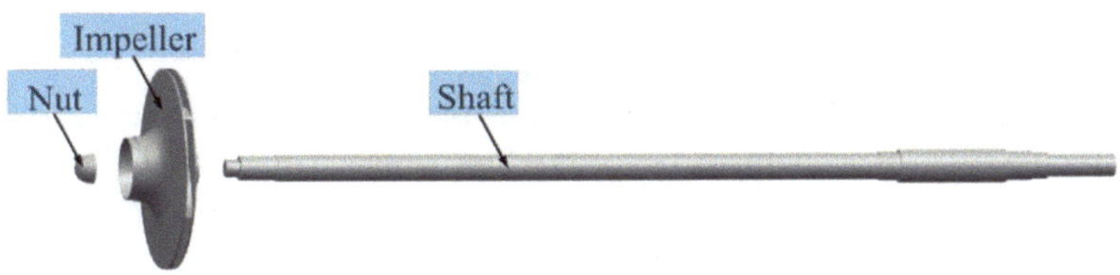

Fig. 9.2 Structural domain of the molten salt pump

only the impeller and shaft are taken into account. The three-dimensional computational model is established and displayed in Figs. 9.1 and 9.2.

The commercial grid-generation code ANSYS ICEM is used to discretize spatially the entire computational domain of the molten salt pump with unstructured grids, and local areas are refined to ensure the physical phenomena are sufficiently captured. In order to reduce the calculation error caused by the grid number, a grid independence examination of the molten salt pump is carried out by devising five sets of grids with different density. The results are tabulated in Table 9.1.

In Table 9.1, the errors of pump head and efficiency are within 1% as the number of the grids is increased to more than 2 million, therefore, the grid scheme C is

Table 9.1 Grid independence examination

Grid	Grid number	Head /m	Efficiency /%
A	1264593	64.5	77.8
B	1691739	65.4	79.4
C	2247046	65.7	79.7
D	2705654	65.6	79.9
E	3271502	66.1	80.3

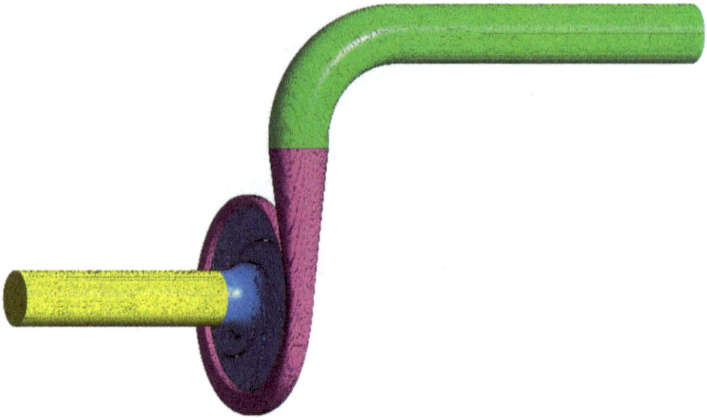

Fig. 9.3 Grid of flow domain

selected for subsequent calculation in consideration of the computational accuracy and economy. The grid deployment for the flow domain is illustrated in Fig. 9.3.

9.1.2.2 Numerical Strategy

Unsteady flows in the high temperature molten-salt pump are simulated with the commercial code ANSYS CFX. The molten salt with dual elements, $NaNO_3$ with the percentage of 60% and KNO_3 with the percentage of 40%, is chosen as the flow medium. Such a medium is used primarily in solar photovoltaic power generation [21]. The physical properties of the medium are displayed in Table 9.2. In compatible with unsteady flow characteristics of the molten-salt pump, the time interval between neighboring simulation steps is set to 0.000574715 s, which means the impeller rotates 5° in each time step. In consideration of both accuracy and cost, nine cycles of impeller rotation are covered.

Uniform velocity distribution is set as the inlet boundary condition, and constant static pressure is set as the outlet boundary condition. No-slip boundary conditions are defined for all solid walls. Standard wall function is utilized in near-wall flow regions. The standard k-ε turbulence model is adopted here.

Table 9.2 Physical properties of working media

Media	Density / (kg/m^3)	Viscosity / (Pa . s)	Temperature / °C	Specific heat capacity J/(kg. K)
Molten salt	1804	0.00147	450	1520

Table 9.3 Comparison between numerical and test result

	Flow rate / (m³/h)	Head /	Efficiency /%
Simulation results	200	65.7	79.4
Experimental results	200	64.6	77.2
Relative error (%)	0	1.7	2.8

9.1.3 Flow Characteristics of the Pump

The performance parameters of the molten-salt pump are obtained through numerically obtained velocity and static pressure data at the inlet and outlet of the computational domain. Numerical results are compared with the test performance of the pump to validate the numerical scheme. The comparison is explained in Table 9.3. The error between the simulation and the test result is less than 3% under the design condition, and the deviation is acceptable. Meanwhile, the reliability of the numerical scheme is proved.

The whole rotation cycle of the impeller is divided into six periods, each is characterized by an identical blade passage. Therefore, the results explained subsequently are represented by three transient phases, P_1, P_2, and P_3, evenly distributed in the one-sixth cycle of the impeller rotation. And the results corresponding to the ninth cycle of impeller rotation are analyzed.

9.1.3.1 Pressure Distribution

Static pressure distributions at P_1, P_2, and P_3 are displayed in Fig. 9.4. Overall, the distribution trend of static pressure in the pump at the P_1, P_2, and P_3 phases is similar. The static pressure increases from impeller inlet to the outlet gradually and the symmetry of static pressure distribution with respect to individual blades is well maintained. Distinct imparity is found between the three phases near the volute tongue. As the relative position between the blade outlet edge and the volute tongue

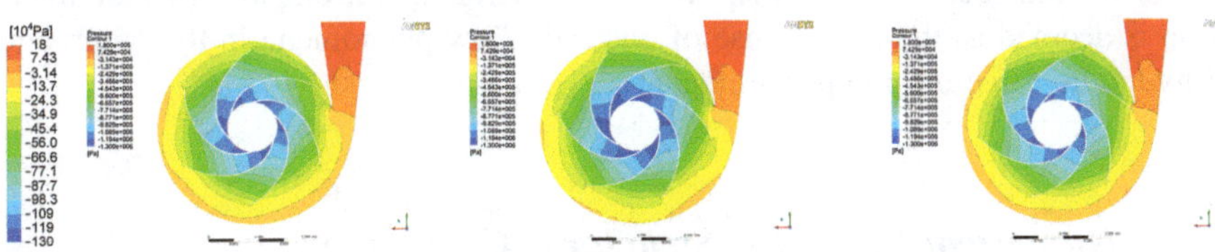

(a)P_1 (b)P_2 (c)P_3

Fig. 9.4 Pressure distribution at different phases

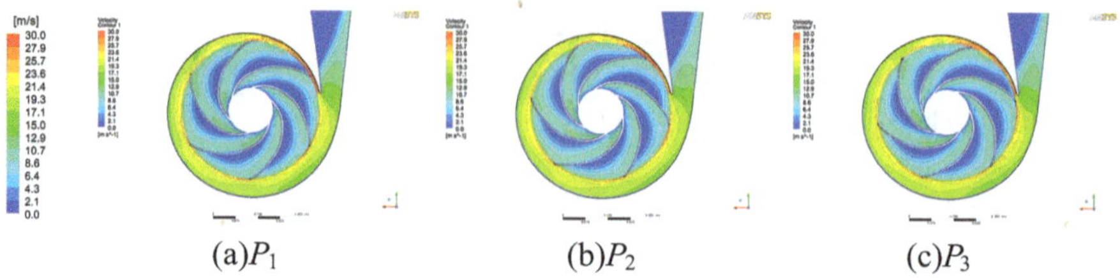

Fig. 9.5 Velocity distribution at different phases

varies, the pressure distribution pattern changes significantly. The periodic rotor-stator interaction arouses a complex pressure distribution over the medium wetted component surfaces.

9.1.3.2 Velocity Distribution

Cross-sectional relative velocity distributions at P_1, P_2, and P_3 phases are displayed in Fig. 9.5. From an overall perspective, the velocity distributions at the three phases are similar. In each impeller passage, low-velocity elements are produced near the pressure surface. The velocity difference between the pressure surface and the suction surface is remarkable. As the relative position between the impeller and the volute tongue varies, the circumferential velocity distribution experiences a periodic variation at impeller outlet. Such a periodic variation influences essentially the hydraulic forces exerted on the impeller and the volute. Provided that the turbulent fluctuations are taken into account, the characteristic fluctuation frequencies can be extracted from transient flow data. Regarding the correlation between the flow-excited frequencies and the vibration frequencies associated with the pump components, they are not necessarily compatible. In this context, multiple factors such as the material, structure, restraint manners of the pump components have to be considered comprehensively. In general, the strong rotor-stator interaction is deemed as the main cause of unsteady flow phenomena in the pump and the hydraulic force on pump components.

9.1.4 Numerical Setup for Structural Calculation

9.1.4.1 Computational Grids

The pump rotor is represented by the combination of the impeller and pump shaft. With the commercial code of ANSYS Workbench, the corresponding structure

Fig. 9.6 Grid of structure domain

Table 9.4 Physical properties of the rotor

Physical quantity	Magnitude
Density /(kg/m^3)	7720
Elastic modulus /GPa	219
Poisson ratio	0.281
Thermal conductivity /(W/(mk))	44
Thermal expansion coefficient /(1/K)	1.02×10^{-5}
Yield strength /MPa	685

computation domain is discretized with unstructured grids. Grid number of the solid part is 236960. The grid deployment of the solid part is shown in Fig. 9.6.

All rotor components are made of martensitic stainless steel ZG1Cr13, whose physical properties are shown in Table 9.4.

9.1.4.2 Load and Constraint Conditions

During the operation of the pump, the rotor is exposed to the effect of gravity, the centrifugal force and fluid pressure. The centrifugal force is imposed by applying the angular velocity to the impeller, the fluid pressure load is transplanted from previously obtained static pressure distribution over the solid surface through the fluid-solid coupling interface. Additionally, cylindrical supports are applied on the upper bearings for structural analysis.

9.1.5 Temperature Distribution in the Rotor

The medium temperature obtained from the flow analysis is imposed on the impeller surface via the interface. Convective heat transfer boundary conditions are set on the bearing surface to ensure the continuous and stable operation of the pump. Convective heat transfer coefficient is 65 W/ (m^{2}k). Only the temperature distribution at P_1 phase is given in Fig. 9.7 in view of the similarity of temperature distributions at different phases.

Fig. 9.7 Temperature distribution over the rotor

In Fig. 9.7, From the impeller to the upper end of the shaft, the temperature decreases from 450 to 22 °C. The impeller is completely immersed in the high temperature molten salt during operation, which causes the temperatures of impeller and medium to be identical. The heat is transferred upwards through the pump shaft by heat conduction and then taken away by the cooling device at the end of the pump shaft. The temperature gradients at both ends of the rotor are high, which produces thermal stress and affects the strength of the whole structure. Therefore, temperature difference should be taken into account in pump design and manufacture.

9.1.6 Strength Analysis of the Rotor

The strength criteria include Tresca criterion and von Mises criterion. With the Tresca criterion, the maximum shear stress of the structure serves as the primary factor that leads to material yield failure. While with the von Mises criterion, the main factor causing the material yield failure is the shape change ratio of the material. Both criteria are widely used in engineering. In some cases, the Tresca criterion is apparently conservative, and the von Mises criterion is deemed as being in good agreement with the practical situation. Therefore, the von Mises criterion is used here.

Von Mises equivalent stress criterion follows the fourth strength theory of the material mechanics, and the equivalent stress can be obtained by:

$$\sigma_e = \sqrt{\frac{(\sigma_1 - \sigma_2)^2 + (\sigma_2 - \sigma_3)^2 + (\sigma_3 - \sigma_1)^2}{2}} \tag{9.1}$$

Generally, if the material meets the requirements depends on whether $\sigma_e < \sigma_s$, and σ_s is the material yield strength. In practical applications, a yield safety factor n_s is needed, then the allowable stress $[\sigma]$ of the material is obtained with:

$$[\sigma] = \sigma_s / n_s \tag{9.2}$$

and then the criterion takes the form:

$$\sigma_e < [\sigma] \tag{9.3}$$

Equivalent effect is calculated by:

$$\varepsilon_e = \frac{1}{1+\nu'}\sqrt{\frac{(\varepsilon_1-\varepsilon_2)^2+(\varepsilon_2-\varepsilon_3)^2+(\varepsilon_3-\varepsilon_1)^2}{2}} \tag{9.4}$$

where ν' denotes the Poisson ratio of material.

9.1.6.1 Stress Analysis of Rotor System

The equivalent stress distributions at the phases of P_1, P_2, and P_3 are shown in Fig. 9.8. The maximum equivalent stress increases from P_1, P_2, to P_3 phases and the corresponding stress values are 236, 268 and 284 MPa, respectively. The overall

(a) P_1

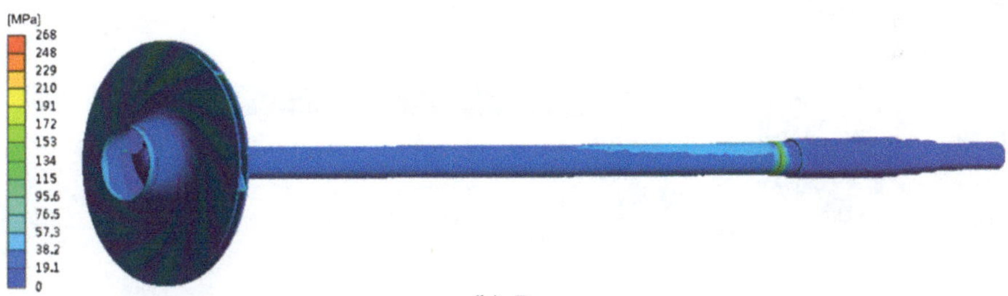

(b) P_2

(c) P_3

Fig. 9.8 Equivalent stress distribution of the rotor

stress distributions at the three phases are similar, and the maximum stress appears on the surface of the lower bearings. The surface of the bearings suffers from the stress caused by the deflection of the pump shaft and impeller rotation. Meanwhile, material thermal expansion triggered by remarkable temperature difference also results in high thermal stress as the restraint is imposed. By contrast, the stress distribution over the impeller is symmetrical. The stress of different impeller passages is uniformly concentrated on blade pressure surface, and the equivalent stress on the pressure surface is larger than that on the back surface. The effect of impeller rotation phase on the pattern of equivalent stress distribution is insignificant.

9.1.6.2 Deformation Analysis of the Rotor

The deformation of the rotor at different phases is illustrated in Fig. 9.9. It is seen that the deformation at P_1 and P_3 phases is more serious that that at P_2 phase. The

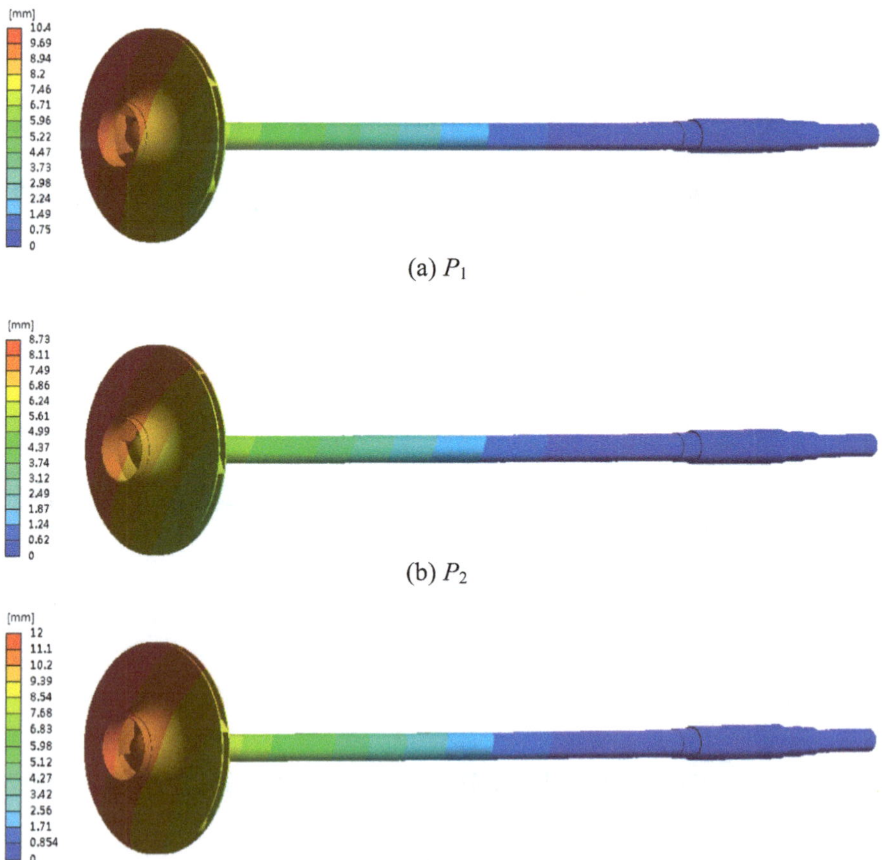

(a) P_1

(b) P_2

(c) P_3

Fig. 9.9 Deformation distribution of the rotor

displacement and deformation in radial direction are caused by thermal stress and the stress generated with fluid-structure coupling. The largest deformation occurs in the impeller, and the deformation decreases consistently upwards along the pump shaft.

9.1.6.3 Strength Verification

According to the calculation result, the maximum equivalent stress of the rotor is 284 MPa, which is less than the yield stress of martensite stainless steel of 685 MPa, as shown in Table 9.4. Then the safety factor of the rotor calculated through Eq. (9.2) is 2.41, providing a sufficient space for safety design.

9.1.7 Modal Analysis

9.1.7.1 Modal Analysis Without Consideration of Prestress

Natural frequencies and amplitudes of the first six orders of the rotor under the condition without prestress are listed in Tables 9.5 and 9.6. The first order of the rotor is similar to the second order in terms of frequency and the same trend appears for the fourth and fifth orders. The natural frequency of the rotor increases consistently from the first to sixth orders. From the fifth to sixth orders, a drastic rise in frequency is evident.

The maximum amplitude of the rotor under the condition of no prestress is 12.69 mm, which appears at the third-order mode. Similar to the natural frequency, the amplitudes of the first order and the second order are close to each other, and the amplitudes of the fourth and the fifth order are equivalent.

9.1.7.2 Prestress Modal Analysis

With the consideration of prestress, natural frequencies and amplitudes of the first six orders of the rotor are shown in Tables 9.7 and 9.8. The phase effect is

Table 9.5 Natural frequency of the rotor without prestress (Hz)

Order	1st	2nd	3rd	4th	5th	6th
Frequency	11.07	11.08	72.84	90.16	90.17	245.38

Table 9.6 Amplitude of the rotor without prestress (mm)

Order	1st	2nd	3rd	4th	5th	6th
Amplitude	6.80	6.80	12.69	7.59	7.59	11.02

Table 9.7 Natural frequency of the rotor with prestress (Hz)

Phase	1	2	3	4	5	6
P_1	11.52	11.52	73.26	90.47	90.47	245.63
P_2	11.53	11.54	73.11	90.46	90.47	245.62
P_3	11.51	11.52	73.28	90.46	90.47	245.60

Table 9.8 Amplitude of the rotor with prestress (mm)

Phase	1	2	3	4	5	6
P_1	6.80	6.80	12.71	7.59	7.59	10.93
P_2	6.80	6.80	12.70	7.59	7.59	10.83
P_3	6.80	6.80	12.71	7.59	7.59	11.12

considered as well. The difference of natural frequency is not obvious at the same order and different phases, with an exception of the third-order mode. The amplitude deviation between the first five orders of the rotor is slight, and the maximum amplitude difference of the sixth-order is only 3%. In this context, the influence of the phase angle on the vibration amplitude is negligible.

With a comparison of natural frequency of the rotor between the conditions with prestress and without prestress, it can be found that natural frequency of the rotor is enlarged with the participation of the prestress. Meanwhile, the tendency of the increase of natural frequency attenuates as the order increases. Essentially, the introduction of prestress enhances the structural stiffness of the rotor. Consequently, the natural frequency increases logically. With respect to the amplitude of the rotor under the two conditions, the influence of prestress is not perceivable over the first five orders of the rotor, but manifests at the sixth order.

9.1.7.3 Vibration Mode Analysis

The first six vibration modes of the rotor is shown in Fig. 9.10. The maximum amplitude of the first, second, third and sixth modes appears in the impeller, while the maximum amplitude of the rest modes appears at the middle of the pump shaft. The first and second modes are dominated by the swing deformation of the pump shaft, and the magnitude of deformation decreases gradually upwards along the pump shaft. The third mode is characterized by the torsional deformation of the pump shaft. In this case, the distribution of deformation is symmetrical, and the largest deformation arises at the outer edge of the impeller. The fourth and fifth modes are distinct due to the secondary swing deformation of the pump shaft, and the amplitude of the pump shaft deformation is larger than that of the impeller. The sixth mode bears bending deformations at different directions, and the maximum deformation of the rotor reappears at the out edge of the impeller. In general, the impeller deformation at low-order modes is evident, while pump shaft deformation at high-order modes is severe.

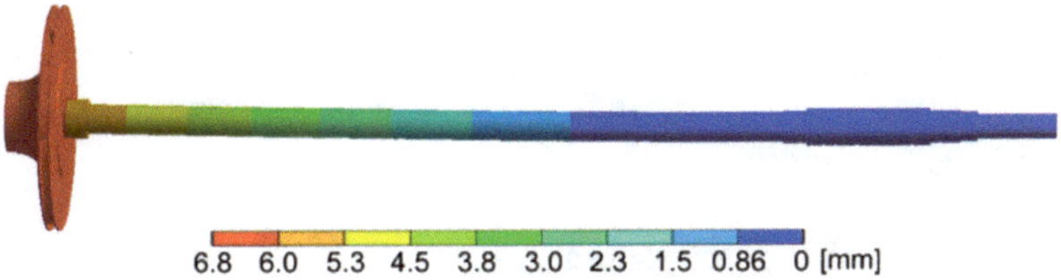

(a) First mode

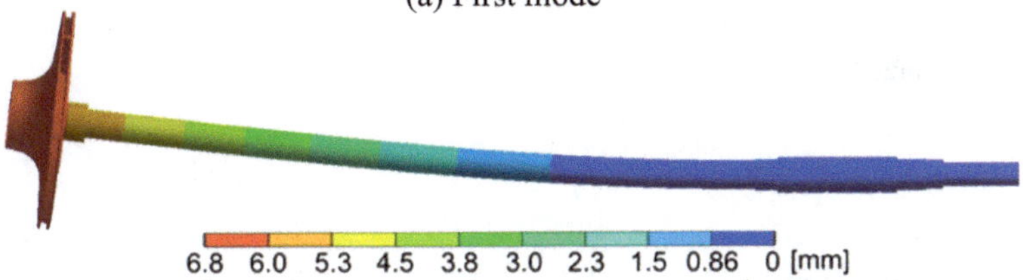

(b) Second mode

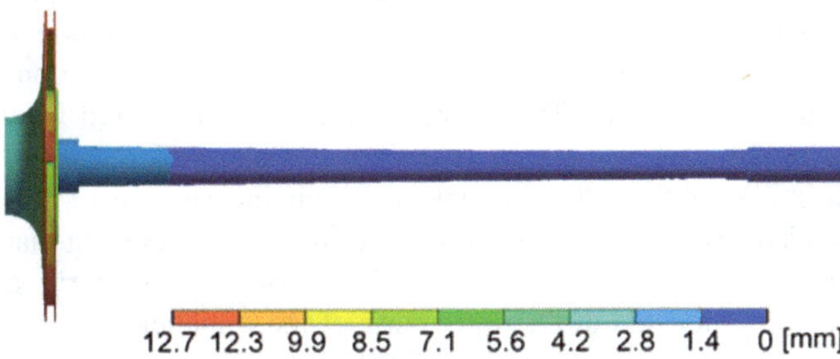

(c) Third mode

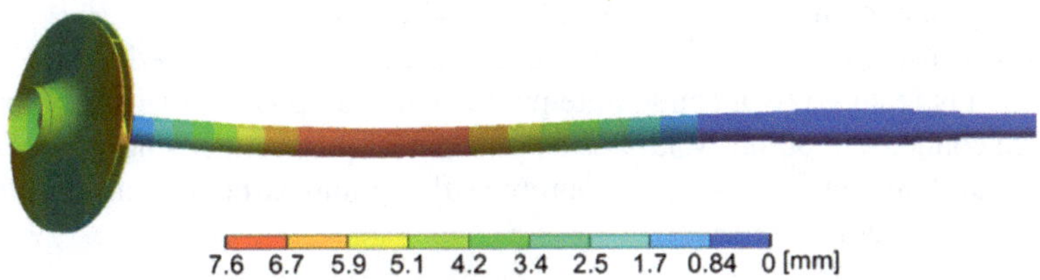

(d) Fourth mode

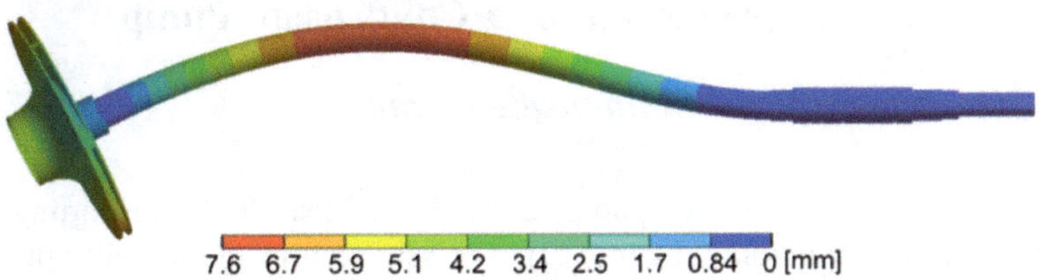

(e) Fifth mode

Fig. 9.10 The first six vibration modes of the rotor

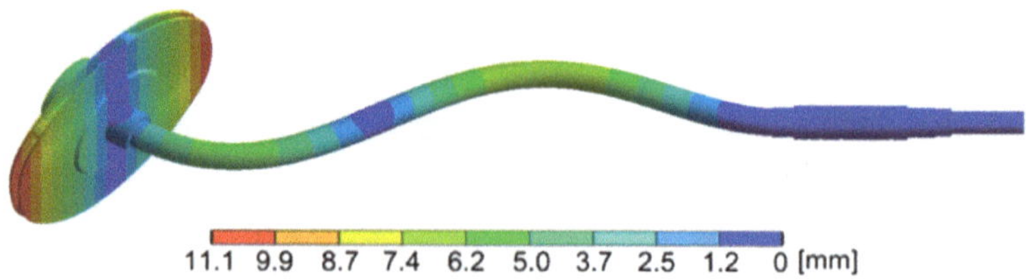

(f) Sixth mode

Fig. 9.10 (continued)

9.1.8 Summary

With this case, the numerical methodology for investigating pump structures as they are exposed to the effect of flow is elucidated. The first step is to carry out unsteady flow simulations, which will enable a description of transient variation of flow parameters as the impeller rotates. It is found that pressure distribution varies periodically with impeller rotation. The connection between flow and solid components is thereby established.

The maximum stress of the rotor system arises on the surface of the lower bearing in this case. This is reasonable in consideration of the supporting manner of the whole rotor. For such a rotor system, the deformation is asymmetric, and the maximum deformation is located at the outer edge of the impeller.

With the calculated natural frequency and amplitude of the rotor system, the relationship between flow and the structural properties is established. This conclusion is valuable and can be extended to a wider range.

A further discussion of the flow-structure interaction is performed for the second case, which is featured by a condensate pump. This pump is a horizontally installed two-stage condensate pump. With this case, it can be proved that studies of the flow and structural aspects can help to optimize the pump structure or to solve the problem of the pump vibration.

9.2 Structural Improvement of a Condensate Pump

9.2.1 Overview of the Condensate Pump

As stated before, the condensate pump serves as a critical fluid-transporting device in the thermal and hydraulic system [22]. The vibration of the impeller pump is a common phenomenon in industrial applications. The extent of the vibration that is lower than some threshold is acceptable [23]. Nevertheless, resonance is not permitted since the vibration amplitude corresponding to the resonance frequency will be augmented to some unexpected extent, sometimes leading to catastrophic events. Existing knowledge of the pump vibration offers some general suggestions as for how to avoid or to alleviate the pump vibration through a delicate pump design [24]. In many cases, both the pump itself and the operation conditions are specific [25]. Therefore, the generalization of the obtained conclusions of the pump vibration is not an easy task.

The vibration of the pump is induced by three main factors, namely mechanical, system and flow factors. Mechanical factors can be easily identified. Among them, misalignment, broken bearings or bended shaft are typical. For system factors, since the operation of the pump is determined by the system, the configuration of the system components such as bends and valves is critical for the practical pump operation flow rate. As the operation flow rate deviates from the design flow rate, the pump vibration is unavoidable, as is essentially related to irregular internal flows of the pump. In this context, the unsteadiness of the flow in the pump is intensified at off-design points and the time-varying hydraulic force imposed on the medium-wetted pump components is enlarged concurrently. Thus far, the flow excitation has been deemed as a very common reason for the pump vibration, frequency analysis proves to be an effective measure of detecting flow-excited pump vibration [26]. In addition, another factor underlying the pump vibration is cavitation [27]. The pump vibration induced by cavitation is featured by multiple high-frequency peaks, which reflect the collapse of cavitation bubbles.

A huge amount of literature has been devoted to the studies of the pump vibration. This justifies the complexity and difficulty in this subject. Meanwhile, practical applications indicate that each pump has its own structural characteristics, so conclusions applying to one pump cannot be transplanted into another. This explains why research efforts have been consumed to carry out diverse experiments on different pumps. Specifically, structural factors such as the configuration of impeller blades, the match between the impeller blade number and vane blade number, the gap between the impeller outer edge and the basic circle of the volute, influence the pump vibration [28].

With computational fluid dynamics, flow characteristics of the pump can be described [29]. Based on numerically obtained static pressure distributions over the surface of the hydraulic components, hydraulically excited forces can be calculated [30]. One of the shortages of the CFD technique is that it cannot describe the vibration propagation in the solid parts of the pump. In this context, finite element method can be used to simulate the stress propagation in the solid parts and to provide data for the modal analysis [31]. This has been explained briefly in Chap. 2.

In practice, various vibration measurement techniques have lent their support to the treatment of pump vibration issues. With instruments such as the piezoelectric transducer, a direct and objective assessment of the pump vibration can be given based on acquired vibration velocity or vibration acceleration [32]. Some studies employed several transducers to measure the vibration acceleration at different monitored points and in multiple directions simultaneously [33]. A comparison of the vibration spectra at different positions facilitates the identification of the vibration source.

With this case, the reason of the vibration of a horizontally installed two-stage condensate pump is sought. Both flow and structural factors that might contribute to the pump vibration are considered. Hence, strategies for suppressing the pump vibration are devised. The proposed strategies are validated using both numerical and test methods.

9.2.2 *Condensate Pump and Vibration Description*

9.2.2.1 Condensate Pump

The condensate pump considered in this case is a two-stage centrifugal pump, as exhibited in Fig. 9.11. As the pump operates, the impeller rotates with a rotational speed, n, of 2985 rpm, the volumetric flow rate, q_{V0}, is 225 m^3/h, and the total pump head of the two stages, H_0, reaches 279 m. The pump and the motor are connected via a John Crane coupling. The weights of the pump and the motor are 900 and 486 kg, respectively. The gravity centers of the pump and the motor are located in the same horizontal plane, which is higher than the ground by 250 mm.

The schematic view of the pump considered is displayed in Fig. 9.12. The impellers of the two stages are different. The impeller diameters of the first- and second-stage impellers are 350 and 325 mm, respectively. The two impellers are installed in a back-to-back manner, which minimizes the hydraulic force in axial direction. The two stages are bridged through a flow passage. In this context, regarding the two-stage impeller pump, the studies on pump vibration and flow-induced pressure fluctuations have rarely been reported hitherto [34].

Fig. 9.11 In-situ images of the condensate pump

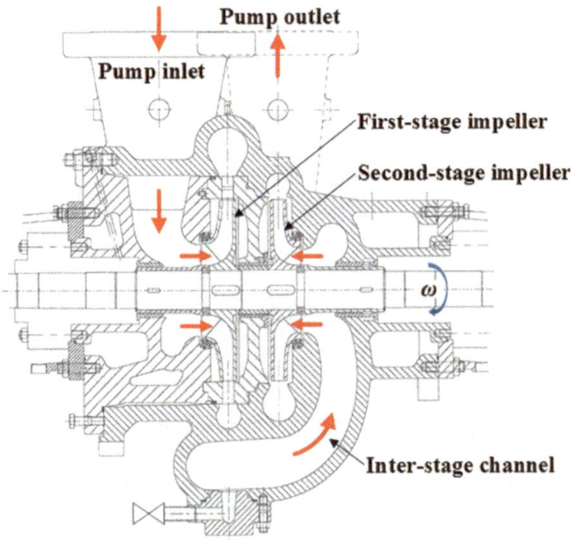

Fig. 9.12 Schematic of the two-stage condensate pump

9.2.2.2 Vibration of the Pump Unit

At the pilot run stage of the condensate pump, with the medium of pure water of 20 °C being transported, an excessive vibration of the pump is detected. The monitored temperature of the bearings arrives at 77 °C. It is thereby conjectured that resonance occurs with the pump unit. In consideration of the medium used, the possibility of cavitation-induced pump vibration is excluded. Meanwhile, all the seals and gaps are examined to check if the vibration is caused by wear or lack of lubricants [35]. These factors are excluded as well. All the cooling equipment is inspected as well. Finally, the pump unit is dismantled to measure the natural frequency of the pump base. The measurement is performed using a CRAS vibration and dynamic signals acquisition system and the maximum frequency of 8 kHz is selected. An impact hammer is used for excitation [36]. It is found that the excitation at high frequencies is considerably weak relative to those low-frequency signals. The results at the frequency range of 0–500 Hz are plotted in Fig. 9.13.

It is seen in Fig. 9.13 that the pump base is featured by low natural frequencies. In horizontal direction, three natural frequencies, close to each other, are dominant. Meanwhile, in vertical direction, the natural frequency of 96.25 Hz is distinct, and the spectra obtained at the two monitored points are nearly identical. It can be judged from Fig. 9.13 that both the mass and stiffness of the pump base are large, and the pump vibration is not aroused by a soft pump base.

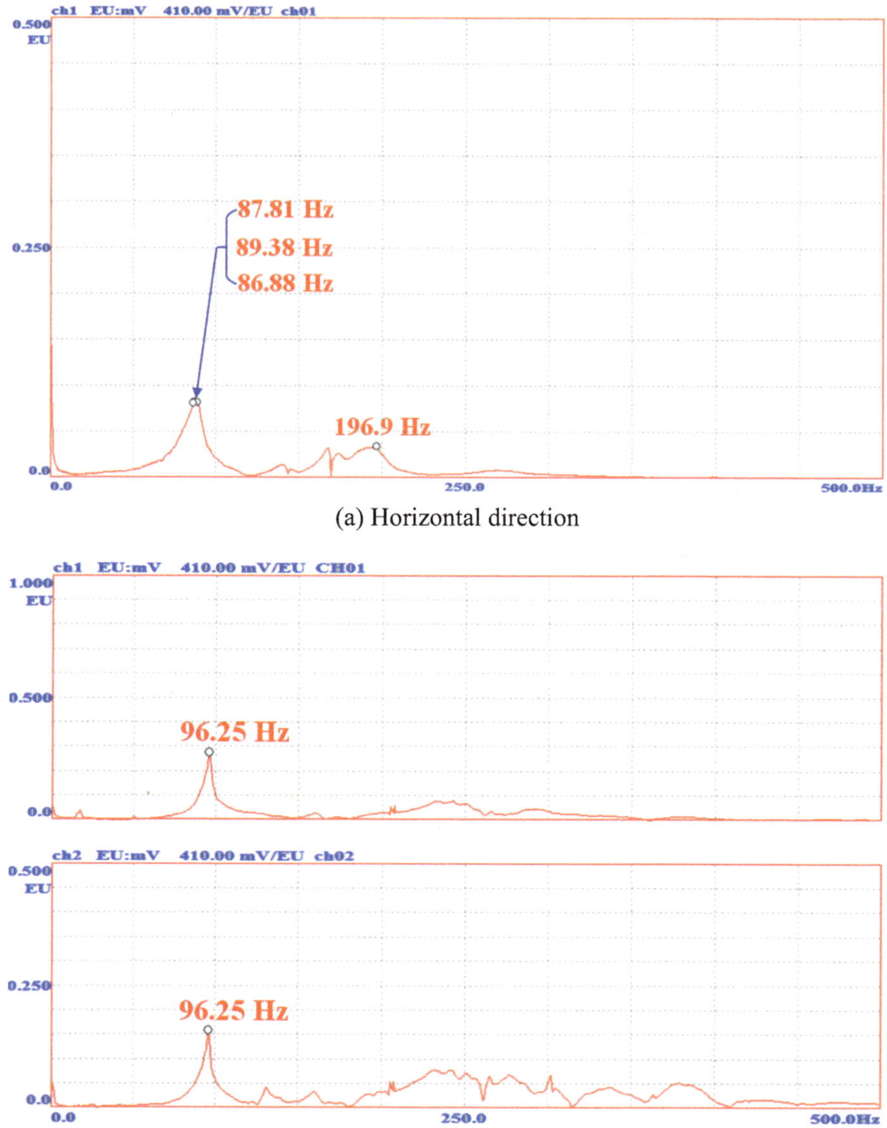

(a) Horizontal direction

(b) Vertical direction

Fig. 9.13 Natural frequencies of the pump base

9.2.3 Flow Simulation for the Pump

To obtain the flow-excited signals, unsteady flow simulation was performed for the whole flow passage in the two-stage condensate pump. Provided that the predominant frequencies in the internal flows are close to the dominant frequencies indicated in Fig. 9.13, the vibration of the pump base is highly related to the flow excitations. Then the cause of the resonance would be determined.

9.2.3.1 Computational Domain

The computational domain of the two-stage condensate pump is exhibited in Fig. 9.14. In consideration of the flow physics and the energy delivered from the impeller blades to the medium, the whole domain is divided into rotating and stationary parts; the former encompasses flow passages of the first-stage and the second-stage impellers, and the latter are composed of the remaining subdomains.

9.2.3.2 Numerical Scheme

The Reynolds-averaged Navier-Stokes (RANS) governing equations are used in the present simulation. The shear stress transport (SST) k-ω turbulence model proved to be capable of capturing multi-scale flow structures produced in the impeller pump [37]. Therefore, the SST k-ω turbulence model is used here to arrive at the closure of the governing equations. The commercial CFD code ANSYS CFX serves as the simulation platform.

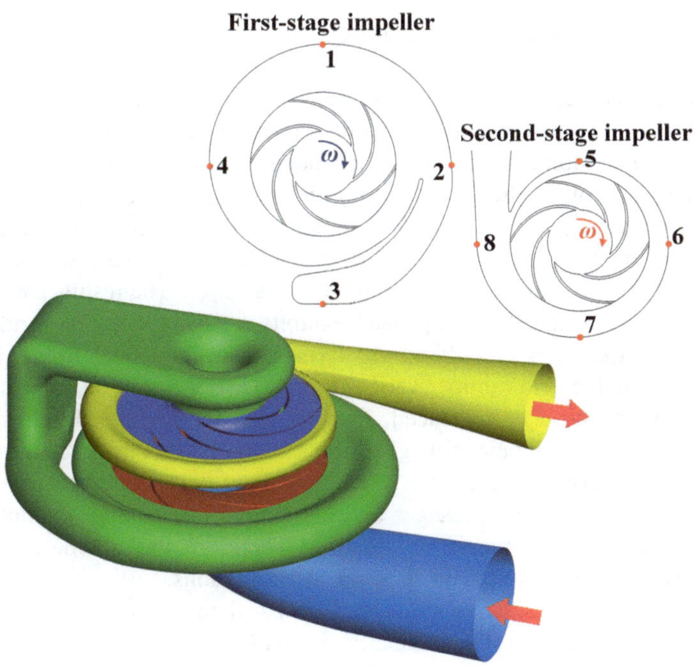

Fig. 9.14 Flow domain of the condensate pump and monitored points

The medium of pure water of 20 °C is used in the simulation and the fluid is assumed to be incompressible. Velocity inlet boundary condition is set at the inlet of the whole flow domain; the velocity magnitude is calculated through the relationship between the velocity magnitude and the flow rate, and uniform velocity distribution over the inlet area is predefined. Pressure outlet boundary condition is set at the outlet of the computational domain. The outlet pressure is kept constant and equal to the local atmospheric pressure. Both the inlet and outlet pipes of the pump are extended to ensure fully developed incident flows and no backflow at the pump outlet section. At all medium-wetted solid walls, no-slip boundary conditions are defined. Standard wall functions are adopted to treat near-wall flows. Identical surface roughness of 0.0125 mm is set for the impeller blade surface, while 0.025 mm for the other solid walls.

To acquire fluctuating signals in the flow field, eight monitored points are deployed on the mid-span plane of the pump casing, as shown in Fig. 9.14. Points 1, 2, 3 and 4 are deployed on the first-stage volute, while Points 5, 6, 7 and 8 are on the second-stage volute. For the unsteady flow simulation, the communication between the rotational subdomains and the adjacent subdomains is accomplished using the moving grid approach. Computational grids are rebuilt continuously with the rotation of the impeller. During the simulation, the time step is set to 5.56×10^{-5} s, which amounts to the time span with which the impeller rotates by 1°. The convergence criteria for each monitored quantity are set to 1×10^{-4}.

9.2.3.3 Validation of the Numerical Scheme

In view of the curved impeller blades and the three-dimensional flow passages bridging the two pump stages, unstructured tetrahedral grids are used to discretize the computational domain. A grid independence examination is performed to quantify the variation of the pump head with the grid number. Six schemes with grid numbers ranging from 1.85 to 7.6 million are devised for the whole computational domain. With identical numerical settings, numerical simulations are conducted for the six schemes at the design flow rate, q_{V0}. The results are plotted in Fig. 9.15. It is seen that the pump head exhibits a clear rise as the grid number increases from 1.85 to 4.37 million, signifying stable numerical results have not been reached. As the grid number varies from 4.37 to 7.6 million, a gradual variation of the pump head is evidenced. In particular, the relative difference of the pump head between each pair of neighboring schemes is less than 1.5% as the grid number exceeds 5.18 million. With these schemes, it can be deemed that flow structures in the pump flow passages and the factors influencing hydraulic losses have been reflected appropriately via the grids. Therefore, the scheme with the grid number of 6.33 million is used in subsequent simulations.

For different flow rates, simulations are performed and the pump head as well as the pump efficiency is calculated. Then a comparison between the numerical and experimental results is illustrated in Fig. 9.16 to evaluate the validity of the numerical scheme constructed.

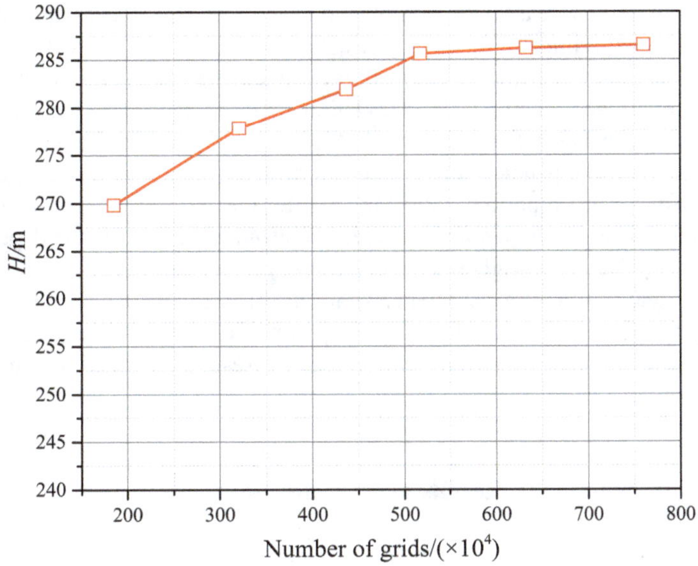

Fig. 9.15 Variation of the pump head with the grid number

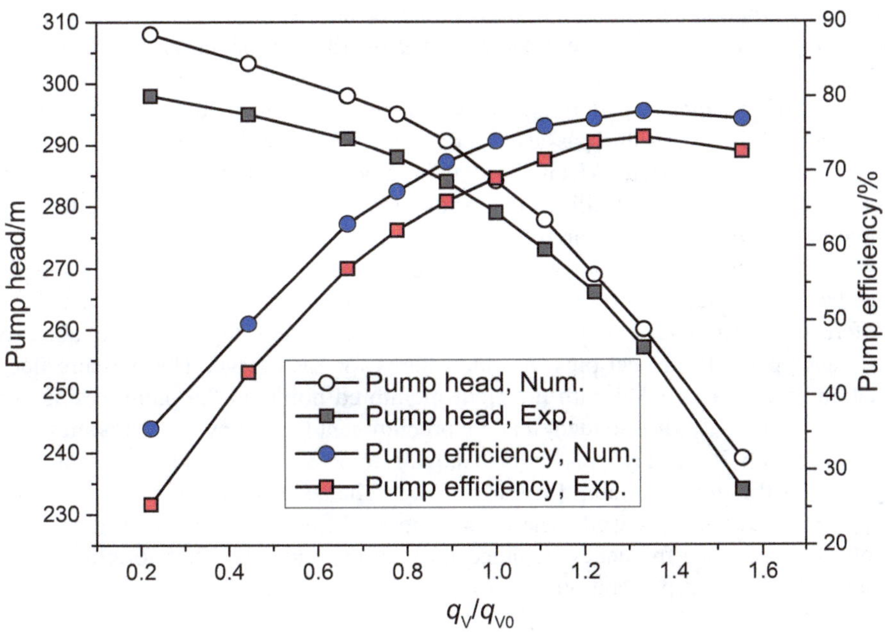

Fig. 9.16 Comparison between numerical (Num.) and experimental (Exp.) results

As the flow rate increases, the pump head drops continuously, as applies to both numerical and experimental data. Numerically obtained pump head is consistently higher than its counterpart, as stems from the inadequate consideration of hydraulic losses in numerical modelling. The deviation between the numerical and experimental results reaches its maximum, 4.3%, at $0.2q_{V0}$, while the minimum gap between the two pump head curves appears at $1.1q_{V0}$. In this context, the volumetric loss plays an important role. Meanwhile, the two curves representing the pump efficiency are in apparent agreement. The numerically obtained pump efficiency is overwhelming higher than corresponding experimental results. As the mechanical and volumetric efficiencies are taken into account, the maximum deviation between the numerically obtained pump efficiency and the practical pump efficiency is lower than 3%. Thus, the numerical results obtained are reliable for further analysis.

9.2.3.4 Flow Simulation Results and Discussion

The operation stability of the pump can be evaluated through flow parameter distributions. The distributions of the total pressure and flow velocity at q_{V0} are shown in Fig. 9.17. As shown in Fig. 9.17a that the total pressure increases gradually from the pump inlet to the pump outlet. After receiving the energy imparted by the first-stage impeller, the medium shows an increase in the total pressure. In the inter-stage flow passage, the total pressure distribution is fairly uniform. The rotation of the second-stage impeller gives rise to a further elevation of the total pressure of the medium. It is noticeable in Fig. 9.17b that both the pump inlet passage and the inter-stage passage are dominated by low velocity, as is related to the spacious flow area. Meanwhile, the velocity distribution patterns in the inter-stage flow passage facilitate the production of hydraulic losses. Nevertheless, there is no drastic velocity variation or high velocity gradients in the flow passage.

The relationship between the flow-induced pump vibration and pressure signals has been consolidated in previous work [38]. For the pump considered here, the difference of pump head between the two rigidly connected stages is remarkable, so it is anticipated to inspect pressure fluctuations for each stage. The pressure fluctuation spectra associated with the eight monitored points in the pump casing are plotted in Fig. 9.18. It is evident that the predominant frequency at each point is the blade passing frequency, f_{BPF}, which equals to 248.75 Hz. Such a situation is typical for the impeller pump [39]. Meanwhile, the second and third harmonics of f_{BPF} are remarkable as well. These characteristic frequencies deviate apparently from the resonance frequency obtained through experiments. Therefore, it is concluded that the pump vibration is not caused by hydraulic excitations.

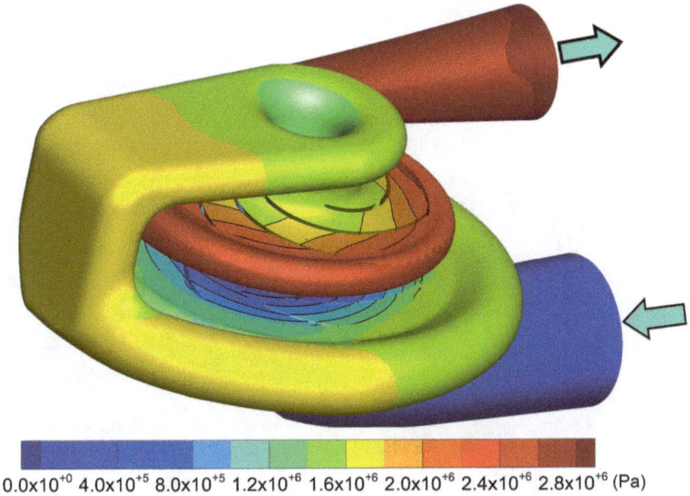

(a) Total pressure distribution

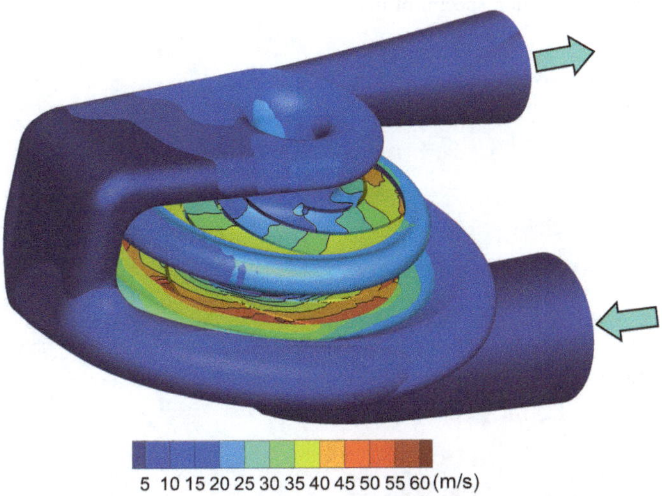

(b) Velocity distribution

Fig. 9.17 Flow parameter distributions in the condensate pump

9.2.4 *Motor Test*

A test is performed to examine the operation of the motor. The motor and the pump were disconnected; thus the motor is operated under no-load conditions. As the rotational speed of the motor approaches 2985 rpm, the vibration velocity at the drive end of the motor is measured to be 2.0 mm/s. Meanwhile, the vibration of the

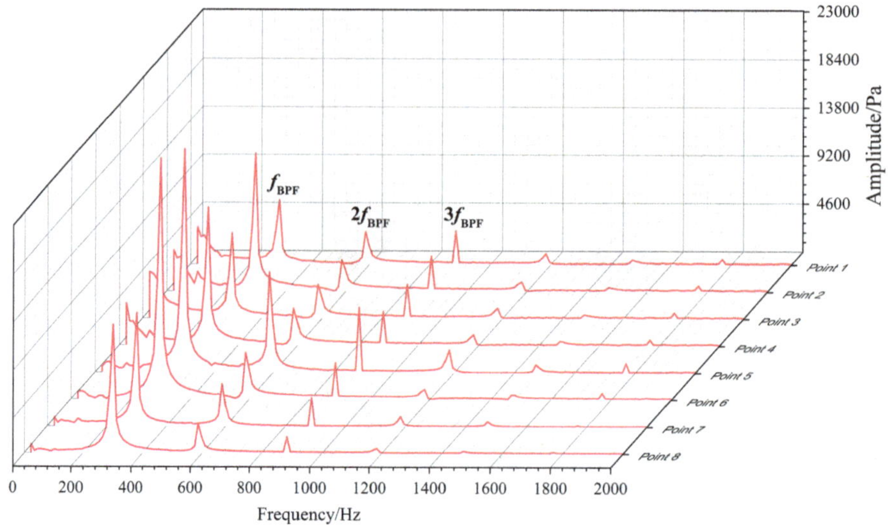

Fig. 9.18 Pressure fluctuation spectra of the monitored points

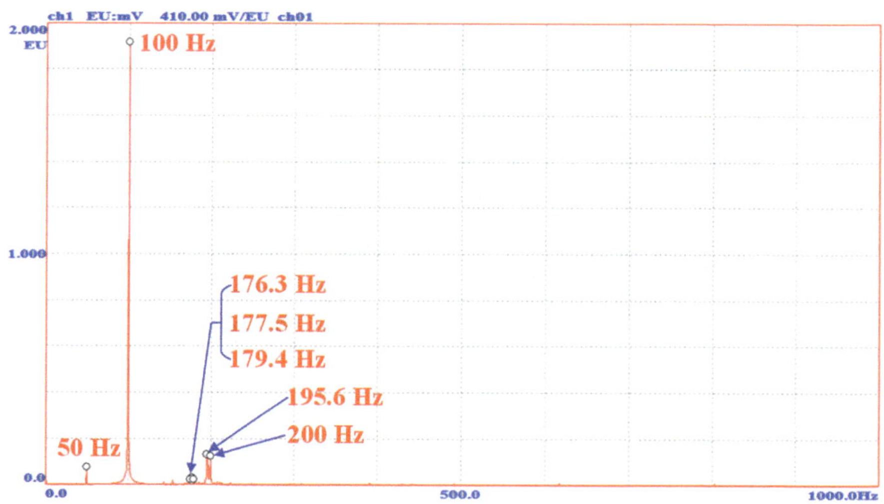

Fig. 9.19 Frequency spectrum of the motor under no-load condition

motor is quantified with the CRAS vibration and dynamic signals acquisition system and the result is plotted in Fig. 9.19. It is seen that there is a distinct peak corresponding to the frequency of 100 Hz. It is thereby deduced that the pump vibration might be aroused by the resonance between the pump base and the motor.

Regarding the motor, it is not easy to alter its natural frequency or vibration characteristics since the electromagnetic components of the motor are integrated and the interaction of these components are difficult to describe and optimize. Therefore, the improvement of the motor supporting or the pump base turns out to be the most feasible measure.

9.2.5 Modal Analysis of the Original Pump Base

Prior to reconstructing the pump base, the natural frequencies of the original pump base are analyzed using finite element method. It is anticipated to find the vibration patterns corresponding to the resonance frequencies. Hence, the improvement will be carried out specifically.

9.2.5.1 Geometric Model

The geometric model of the whole pump base is displayed in Fig. 9.20. During the FEM analysis, the mass of the motor and the pump is exerted onto the pump base. According to practical operation conditions, the foundation bolts are used to fix the pump base, so all the degrees of freedom of all the foundation bolt holes is set to zero. With the commercial code of ANSYS Workbench, the solid computational domain, discretized with unstructured grids, is treated numerically [40]. The pump base is made of Q235 steel. The initial internal stress of the pump base is not considered in the FEM simulation, as is in accordance with practical applications, which employ aging treatment to eliminate internal stress. The effect of gravity is taken into account.

Modal analysis principles explained in Chap. 2 are used here. The Block-Lanczos method embedded in ANSYS Workbench is used to obtain the natural frequency and vibration patterns of the pump base.

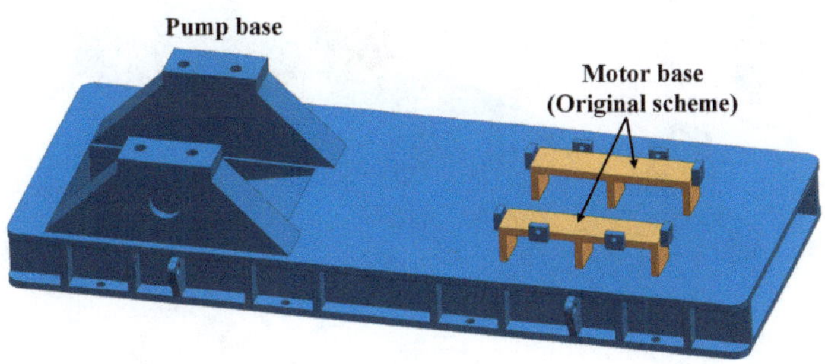

Fig. 9.20 Geometric model of the pump base

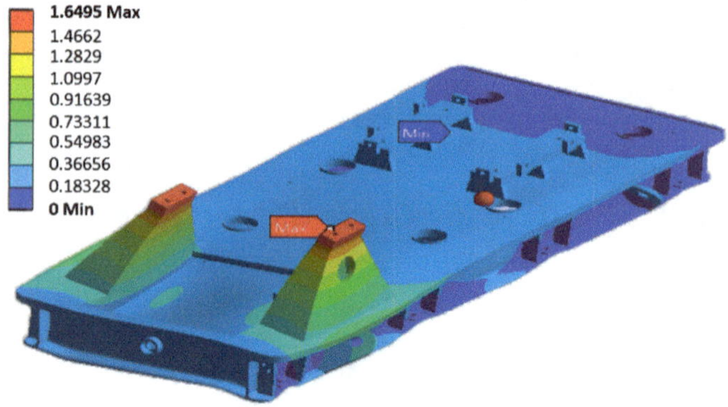

(a) First-order mode

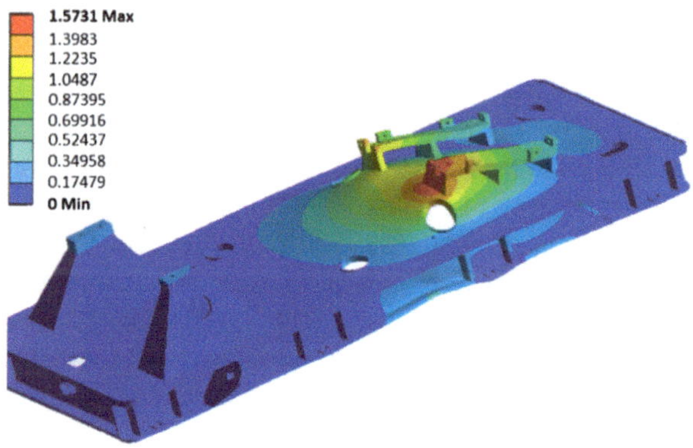

(b) Second-order mode

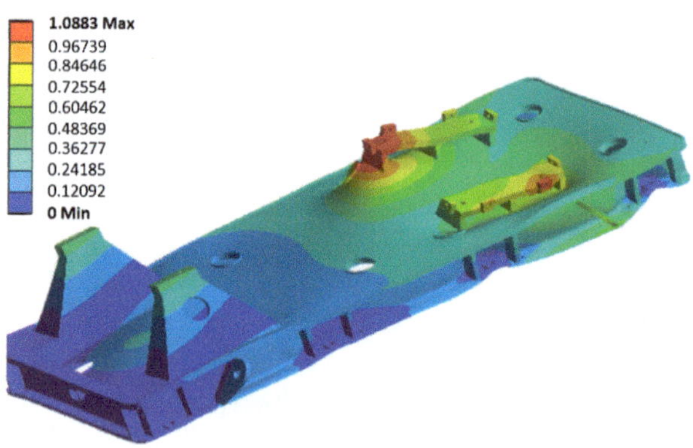

(c) Third-order mode

Fig. 9.21 Vibration patterns of the original pump base (unit: mm)

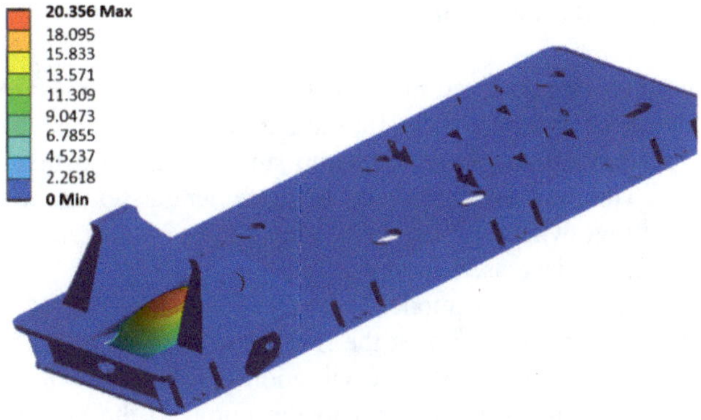

(d) Fourth-order mode

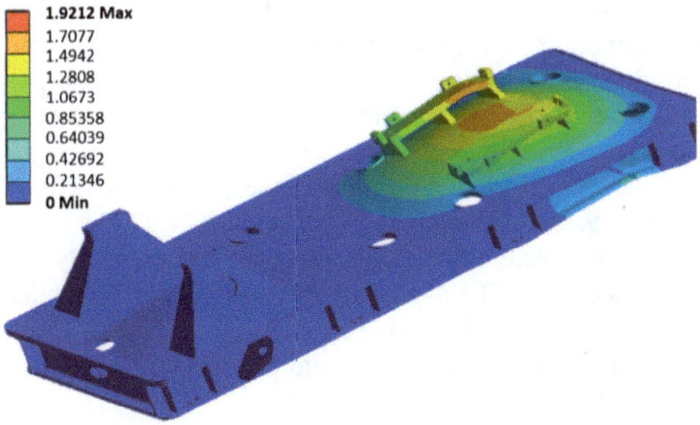

(e) Fifth-order mode

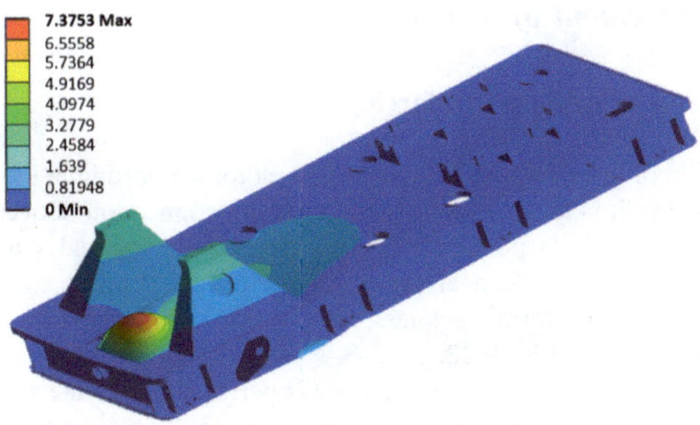

(f) Sixth-order mode

Fig. 9.21 (continued)

9.2.5.2 Results for the Original Pump Base

The first six-order modes obtained numerically are illustrated in Fig. 9.21. For the first-order mode, as shown in Fig. 9.21a, the natural frequency is 85.584 Hz. It is seen that severe vibration occurs at the pump supporting and exhibits the state of horizontal swing. The maximum vibration displacement arises in the upper part of the pump supporting, where the mass of the pump is concentrated. Then the vibration displacement decreases downwards.

Regarding the second-order mode, the vibration is intensified at the motor supporting and the vertical vibration of the side adjacent to the pump is remarkable, as shown in Fig. 9.21b. Meanwhile, the vibration displacement of the motor supporting decreases from the side adjacent to the pump to the non-drive end. The natural frequency of the second-order mode is 110.75 Hz. The natural frequency of 117.72 Hz is associated with the third-order mode. As shown in Fig. 9.21c, the vibration of the motor supporting is apparent, as is similar to that shown in Fig. 9.21b. The maximum vibration displacement appears at the pump side as well. It can be inferred from the natural frequencies and the vibration patterns that the resonance stems from the second- and third-order modes. Therefore, it is critical to alter the frequencies corresponding to the second- and third-order modes.

The fourth-order mode is featured by the vertical vibration of the pump base, as is evidenced in Fig. 9.21d. The natural frequency is 124.97 Hz. For the fifth-order mode, the vibration is severe at the motor supporting; particularly, the vibration of the non-drive end is drastic. The natural frequency is 135.02 Hz, much higher than the resonance frequencies. With respect to the sixth-order mode, the natural frequency is 215.07 Hz, and the vertical vibration of the part of the pump base right underneath the pump is predominant.

9.2.6 *Improvement of Motor Supporting*

9.2.6.1 Scheme of Adding Reinforcing Ribs

In consideration of the vibration patterns of the motor supporting at the second- and third-order modes, it is anticipated to strengthen the connection between the motor supporting and the whole pump base; concurrently, the vertical vibration of the motor supporting at the side near the pump should be suppressed. To this end, reinforcing ribs with uniform thickness of 50.0 mm are added to the motor supporting, as illustrated in Fig. 9.22.

Finite element analysis of the new scheme is performed with the same numerical settings as those used for the original scheme. The vibration frequencies associated with the six modes are plotted in Fig. 9.23. It is evident that for the second- and

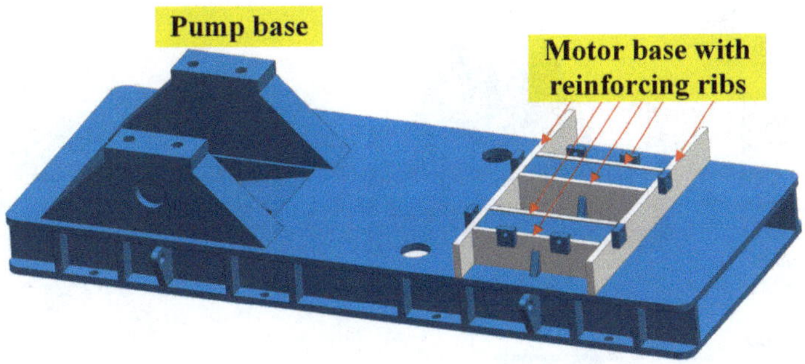

Fig. 9.22 Motor supporting with added reinforcing ribs

third-order modes, the resonance frequency of 100 Hz has been avoided after the ribs are installed. The natural frequency of the pump base increases consistently from the first- to sixth-order modes. For the second-, third- and fourth-order modes, the response to the external excitation is similar. From the fifth- to sixth-order modes, a drastic elevation in the natural frequency is apparent.

The second- and third-order modes are particularly emphasized in the present study. As shown in Fig. 9.24a, the position where drastic vibration occurs shifts from the drive end to the non-drive end of the motor supporting in comparison with the vibration pattern displayed in Fig. 9.21b. Meanwhile, the vibration of the pump base is alleviated and the overall vibration displacement is minimized relative to its counterpart. For the third-order mode, the vibration pattern is shown in Fig. 9.24b. The vibration pattern of the motor supporting is similar to the original scheme, but the resonance frequency has been avoided.

Fig. 9.23 Comparison of natural frequencies between the original pump base and the pump base with reinforcing ribs

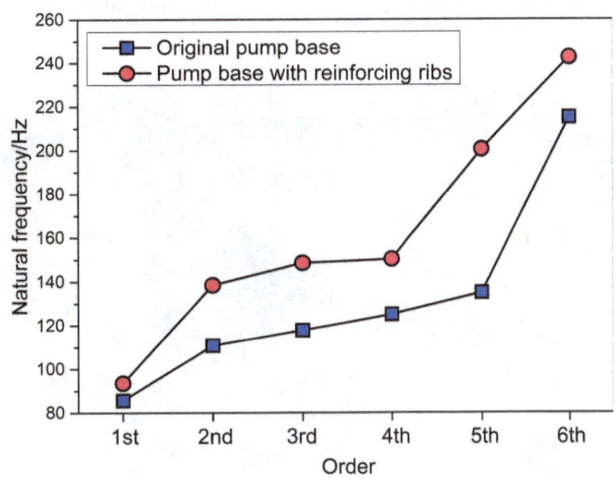

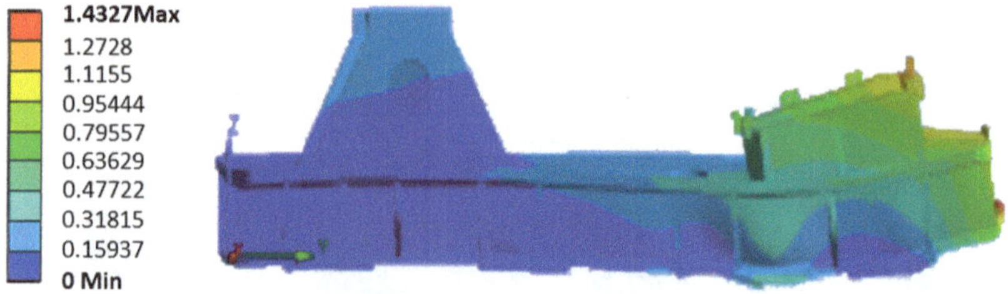

(a) Second-order mode

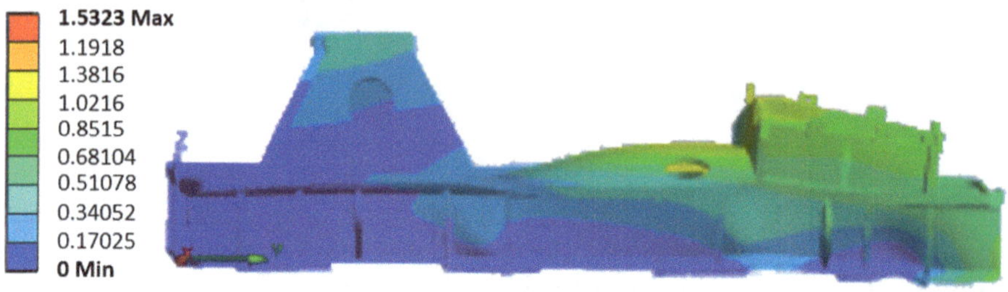

(b) Third-order mode

Fig. 9.24 Second- and third-order modes of the pump base with reinforcing ribs (unit: mm)

9.2.6.2 Experimental Validation

To examine the effect of the improved motor supporting, experiments are carried out. The image of the motor supporting during the welding operation is exhibited in Fig. 9.25, where the dash lines indicate weld joints.

The measurement of the natural frequency of the pump base is conducted using the same method as that employed previously. The results at two monitored points, deployed respectively at the drive end and the non-drive end of the motor

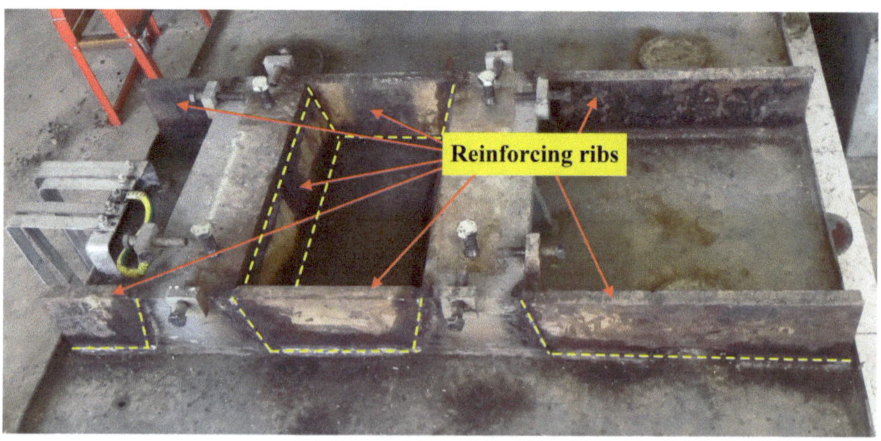

Fig. 9.25 Image of the motor supporting during welding operation

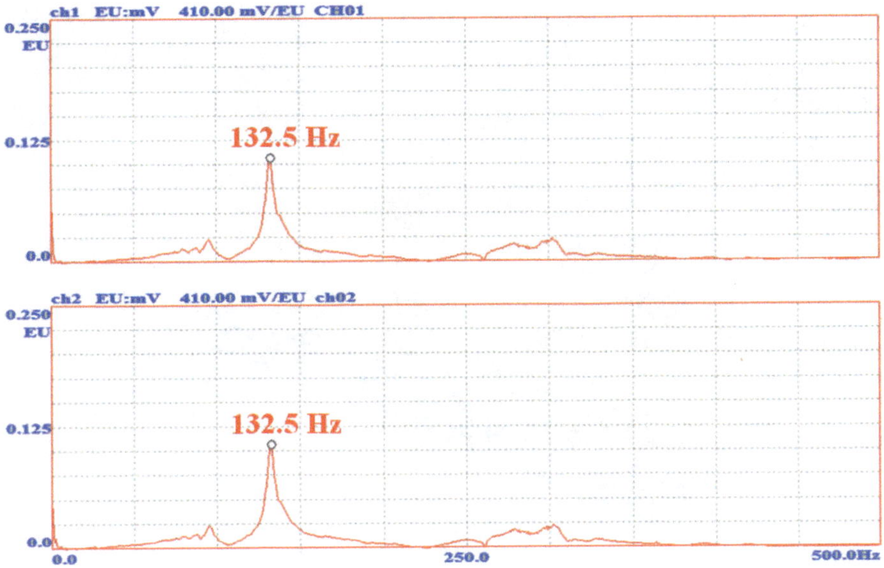

Fig. 9.26 Vibration frequency spectra at two monitored points for the scheme with reinforcing ribs

supporting, are plotted in Fig. 9.26. The dominant frequency is 132.5 Hz, which is considerably higher than the resonance frequency, confirming the positive effect of the reinforcing ribs. Meanwhile, it is seen that the two points exhibit nearly identical frequency spectra. This demonstrates that both the integrity of the motor supporting and the connection between the motor supporting and the whole pump base have been strengthened with the participation of the reinforcing ribs.

9.2.6.3 Secondary Grouting Strategy

Apart from adding reinforcing ribs, the method of secondary grouting is attempted to change the natural frequencies of the pump base. The geometric model of the pump base treated with secondary grouting is displayed in Fig. 9.27. With respect to the secondary grouting, the motor supporting remains the same as that of the original scheme, but the concrete is introduced and the pump base and the concrete are integrated. During the numerical preparation, the freedom of all the surfaces of the concrete are completely restricted. FEM simulations are performed for such a scheme and a modal analysis is implemented in the same manner as that adopted in the above case.

A comparison of the natural frequency between the scheme with secondary grouting and the original scheme is illustrated in Fig. 9.28. The natural frequency increases overall after secondary grouting, as is similar to previously reported results in the vertical pump [41]. Particularly, for the second- and third-order

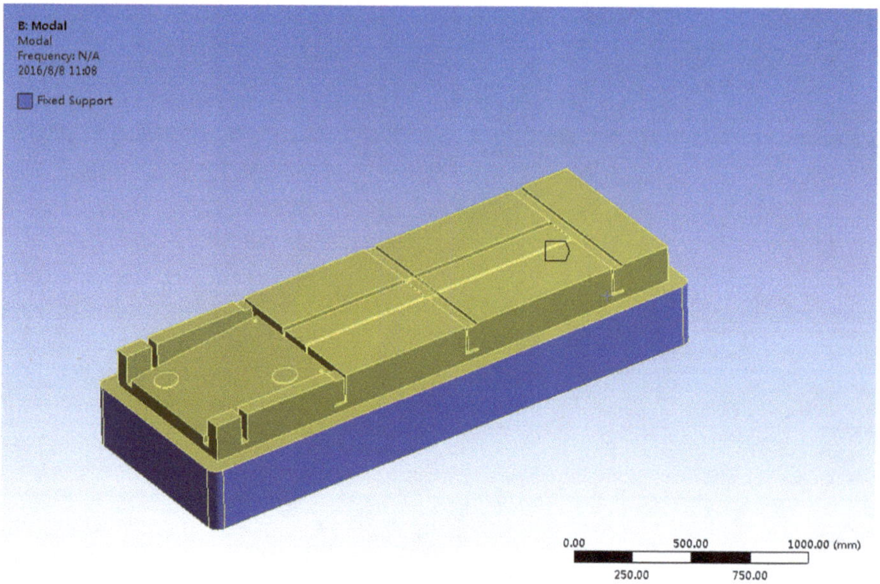

Fig. 9.27 Geometric model of the pump base with secondary grouting

Fig. 9.28 Comparison of the natural frequency between the original scheme and the scheme with secondary grouting

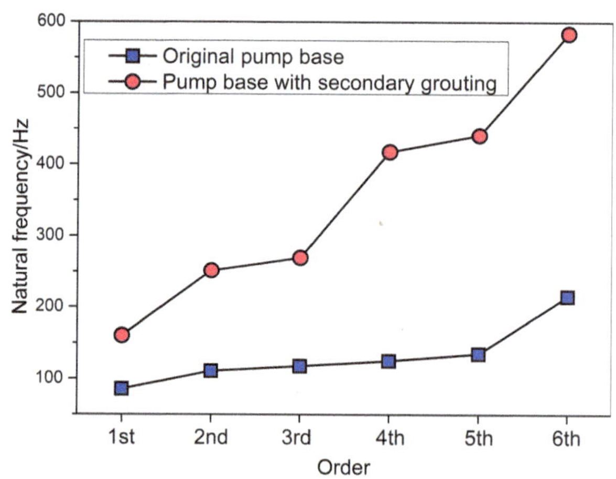

modes, the natural frequencies are considerably high, deviating clearly from the resonance frequency.

Vibration patterns corresponding to the second- and third-order modes are shown in Fig. 9.29. The suppression of vibration is manifested in Fig. 9.29a, where the second-order mode is associated with the vibration of the motor supporting. Such a pattern demonstrates a relatively mitigated vibration in comparison with the original scheme. Meanwhile, the vibration pattern transforms from the vertical

(a) Second-order mode

(b) Third-order mode

Fig. 9.29 Vibration patterns of the scheme of secondary grouting (unit: mm)

vibration of the drive end into the transverse beam vibration of the motor supporting. Regarding the third-order mode, the vibration is apparent in the motor supporting, as is similar to the second-order mode. Essentially, the stiffness of the whole pump base is enhanced due to increased mass. Although the vibration is seemingly concentrated at the motor supporting, the vibration of the whole pump base is mitigated.

In terms of the pump structure, the condensate pump differs clearly from the molten-salt pump. Meanwhile, the focused point with the second case is the resonance problem that has been witnessed in practical engineering. With this case, the function of numerical simulation is manifested. Based on the resonance frequency detected in practice, reasons of resonance are sought. Using CFD, flow-excited pressure fluctuations are simulated and the blade passing frequency and its harmonics are proved to be the dominant frequencies; therefore, the possible contribution of hydraulic factors to the resonance is excluded. A further experiment shows that the excitation frequency of the motor is close to the resonance frequency. A strategy for eliminating resonance is to add reinforcing ribs to the motor supporting. The modal analysis indicates that the natural frequencies corresponding to the second- and third-order modes deviate remarkably from the resonance frequency. The vibration patterns are improved as well. Experimental results also confirm the validity of such a strategy. Another method of secondary grouting is proposed and examined using finite element analysis. The results indicate that the natural frequencies are elevated considerably relative to the scheme featured by reinforcing ribs. Therefore, the resonance problem of the condensate pump is solved.

References

1. Barth DL, Pacheco JE, Kolb WJ, Rush EE. Development of a high-temperature, long-shafted, molten-salt pump for power tower applications. J SolEnergy Eng. 2002;124(2):1059–1080.
2. Hiroshi H, Yoshio Y, Akio S, Yutaka T. Study on design of molten salt solar receivers for beam-down solar concentrator. Sol Energy. 2006;80(10):1255–1262.
3. Chang JC, Yang CH, Yang HH, Hsueh ML, Ho W-Y, Chang JY, Sun IW. Pyridinium molten salts as co-adsorbents in dye-sensitized solar cells. Solar Energy. 2011;85(1):174–179.
4. Wang Z, Naterer GF, Gabriel KS, Secnik E, Gravelsins R, Daggupati V. Thermal design of a solar hydrogen plant with a copper–chlorine cycle and molten salt energy storage. Int J Hydrogen Energy. 2011;36(17):11258–11272.
5. Flueckiger S, Yang Z, Garimella SV. An integrated thermal and mechanical investigation of molten-salt thermocline energy storage. Appl Energy. 2011;88(6):2098–2105.
6. Shao C, Zhou J, Cheng W. Experimental and numerical study of external performance and internal flow of a molten salt pump that transports fluids with different viscosities. Int J Heat Mass Transf. 2015;89:627–640.
7. Kang C, Zhou L, Wang W, Yang M. Influence of axial vane on inner flow and performance of a molten-salt pump, ASME-JSME-KSME. Joint Fluids Eng Conf. 2011;2011:249–255.
8. Childs DW. Fluid-structure interaction forces at pump-impeller-shroud surfaces for rotordynamic calculations. J Vib Acoust Stress Reliab Des. 1989;111(3):216–225.
9. Hu Y, Wang D, Yuan F, Yin J. Numerical study on rotordynamic coefficients of the seal of molten salt pump. Nucl Sci Tech. 2016;27(5):168–178.
10. Fu Q, Yuan S, Zhu R. Calculation and analysis of the critical rotating speed of the rotor system of a centrifugal charging pump in a nuclear power plant. J Eng Thermal Energy Power. 2012;27(5):604–609.
11. Benra F-K, Dohmen HJ. Comparison of pump impeller orbit curves obtained by measurement and FSI simulation. In: ASME 2007 pressure vessels and piping conference, 2007, p. 41–48.
12. Peng GJ, Wang ZW, Yan ZG, Liu RX. Strength analysis of a large centrifugal dredge pump case. Eng Fail Anal. 2009;16(1):321–328.
13. Ashri M, Karuppanan S, Patil S, Ibrahim I. Modal analysis of a centrifugal pump impeller using finite element method. In: MATEC Web of Conferences. EDP Sciences, 2014.
14. Subramaniam L, Sendilvelan S. Modal analysis of a centrifugal pump impeller. Eur J Sci Res. 2012;79(1):5–14.
15. Artal Bartolo E, Cassou-Noguès P, Luengo I, Melle Hernández A. A modal approach for vibration analysis and condition monitoring of a centrifugal pump. Int J Eng Sci Technol. 2011;3(8):321–343.

16. Kang C, Zhang G, Li L, Li Y. Flow and heat transfer in an air-cooling device for a molten-salt pump. In: ASME/JSME/KSME 2015 joint fluids engineering conference, 2015: V001T09A007.
17. Egusquiza E, Valero C, Huang X, Jou E, Guardo A, Rodriguez C. Failure investigation of a large pump-turbine runner. Eng Fail Anal. 2012;23:27–34.
18. Yuan S, Pei J, Yuan J. Numerical investigation on fluid structure interaction considering rotor deformation for a centrifugal pump. Chin J Mech Eng. 2011;24(4):539–545.
19. Tang WZ, Yang L, Zhu W, Zhou YC, Guo JW, Lu C. Numerical simulation of temperature distribution and thermal-stress field in a turbine blade with multilayer-structure TBCs by a fluid-solid coupling method. J Mater Sci Technol. 2016;32(5):452–458.
20. Langthjem MA, Olhoff N. A numerical study of flow-induced noise in a two-dimensional centrifugal pump. Part I. Hydrodynamics. J Fluids Struct. 2004;19(3):349–368.
21. Kearney D, Kelly B, Herrmann U, Cable R, Pacheco J, Mahoney R, Price H, Blake D, Nava P, Potrovitza N. Engineering aspects of a molten salt heat transfer fluid in a trough solar field. Energy. 2004;29(5):861–870.
22. Lancha AM, Serrano M, Briceño DG. Failure analysis of a condensate pump shaft. Eng Fail Anal. 1999;6(6):337–353.
23. Egusquiza E, Valero C, Valentin D, Presas A, Rodriguez CG. Condition monitoring of pump-turbines. New Challenges Meas. 2015;67:151–163.
24. Mustata SC, Dracea D, Tronac AS, Sarbu N, Constantin E. Diagnosis and vibration diminishing in pump operation. Procedia Eng. 2015;100:970–976.
25. Bordoloi DJ, Tiwari R. Identification of suction flow blockages and casing cavitations in centrifugal pumps by optimal support vector machine techniques. J Braz Soc Mech Sci Eng. 2017;39:2957–2968.
26. Osada T, Kawakami T, Yokoi T, Tsujimoto Y. Field study on pump vibration and ISO's new criteria. J Fluids Eng-Trans ASME. 1999;121:798–803.
27. Adamkowski A, Henke A, Lewandowski M. Resonance of torsional vibrations of centrifugal pump shafts due to cavitation erosion of pump impellers. Eng Fail Anal. 2016;70:56–72.
28. Lomakin VO, Chaburko PS, Kuleshova MS. Multi-criteria optimization of the flow of a centrifugal pump on energy and vibroacoustic characteristics. Procedia Eng. 2017;176:476–482.
29. Spence R, Amaral-Teixeira J. Investigation into pressure pulsations in a centrifugal pump using numerical methods supported by industrial tests. Comput Fluids. 2008;37:690–704.
30. Kang C, Li Y. The effect of twin volutes on the flow and radial hydraulic force production in a submersible centrifugal pump. Proc Inst Mech Eng, Part A: J Power Energy. 2015;229 (2):221–237.
31. El-Gazzar DM. Finite element analysis for structural modification and control resonance of a vertical pump. Alexandria Eng J. 2017;56:695–707.
32. Zhang M, Jiang Z, Feng K. Research on variational mode decomposition in rolling bearings fault diagnosis of the multistage centrifugal pump. Mech Syst Signal Process. 2017;93:460–493.
33. Buono D, Siano D, Frosina E, Senatore A. Gerotor pump cavitation monitoring and fault diagnosis using vibration analysis through the employment of auto-regressive-moving-average technique. Simul Model Pract Theory. 2017;71:61–82.
34. Zhang J, Cai S, Li Y, Zhu H, Zhang Y. Visualization study of gas–liquid two-phase flow patterns inside a three-stage rotodynamic multiphase pump. Exp Thermal Fluid Sci. 2016;70:125–138.
35. Norrbin CS, Childs DW, Phillips S. Including housing–casing fluid in a lateral rotordynamics analysis on electric submersible pumps. J Eng Gas Turbines Power. 2017;139(6):062505–1–12.
36. Minette RS, SilvaNeto SF, Vaz LA, Monteiro U. Experimental modal analysis of electrical submersible pumps. Ocean Eng. 2016;124:168–179.

37. Long Y, Zhu RS, Wang DZ, Yin JL, Lin TB. Numerical and experimental investigation on the diffuser optimization of a reactor coolant pump with orthogonal test approach. J Mech Sci Technol. 2016;30(11):4941–4948.
38. Al-Qutubm AM, Khalifa AE, Al-Sulaiman FA. Exploring the effect of V-shaped cut at blade exit of a double volute centrifugal pump. J Pressure Vessel Technol-Trans ASME. 2012;134:021301–1–8.
39. Kumar A, Kumar R. Time-frequency analysis and support vector machine in automatic detection of defect from vibration signal of centrifugal pump. Measurement. 2017;108:119–133.
40. Hsu C-N. Experimental and performance analyses of a turbomolecular pump rotor system. Vacuum. 2015;121:260–273.
41. Tompkins M, Stakenborghs R, Kramer G. Use of FEA software in evaluating pump vibration and potential remedial actions to abate resonance in industrial vertical pump/motor combinations. In: Proceedings of the 2016 24th international conference on nuclear engineering, ICONE24, 26–30 June, 2016, Charlotte, USA.

Kolibri 360 GmbH
Ferdinand-Jühlke-Straße 7
99095 Erfurt, Deutschland
produktsicherheit@kolibri360.de